Getting Started with Electronic Projects

Build thrilling and intricate electronic projects using
LM555, ZigBee, and BeagleBone

Bill Pretty

[PACKT] open source *
PUBLISHING   community experience distilled

BIRMINGHAM - MUMBAI

Getting Started with Electronic Projects

First published: January 2015

Production reference: 1080115

Published by Packt Publishing Ltd.
Livery Place
35 Livery Street
Birmingham B3 2PB, UK.

ISBN 978-1-78355-451-5

www.packtpub.com

Credits

Author
Bill Pretty

Reviewers
Mario Baldini

Chirag Nagpal

Dr. Philip Polstra

Glenn Vander Veer

Commissioning Editor
Amarabha Banerjee

Acquisition Editor
Owen Roberts

Content Development Editor
Pooja Nair

Technical Editors
Vivek Arora

Siddhi Rane

Copy Editor
Vikrant Phadkay

Project Coordinator
Leena Purkait

Proofreaders
Stephen Copestake

Ameesha Green

Indexers
Rekha Nair

Priya Sane

Graphics
Sheetal Aute

Production Coordinator
Shantanu N. Zagade

Cover Work
Shantanu N. Zagade

About the Author

Bill Pretty began his career in electronics in the early '80s with a small telecom start-up company that would eventually become a large multinational. He left that company to pursue a career in commercial aviation in Canada's north. From there, he joined the Ontario Center for Microelectronics, a provincially funded research and development center. After that, a career in the military as a civilian contractor at what was then called the Defense Research Establishment Ottawa. This began a career that was to span the next 25 years, and continues today.

Over the years, he has acquired extensive knowledge in the field of technical security and started his own company in 2010. That company is called William Pretty Security Inc. and provides support, in the form of research and development, to various law enforcement and private security agencies.

He has published and presented a number of white papers on the subject of technical security. He was also a guest presenter for a number of years at the Western Canada Technical Conference, a law enforcement-only conference held every year in western Canada. A selection of these papers is available for download from his website (http://www.williamprettysecurity.com/).

There a number of people I would like to thank, for without their support, this book would never have been completed. My good friends at Packt Publishing for having the patience and trust in me once again. To my partner in life, Donna, who never stopped believing in me. Last but not least, my good friends and fellow code warriors Willie the Mad Scott and Glen the flying Dutchman.

About the Reviewers

Mario Baldini graduated in Computer Science from the Federal University of Santa Catarina and enrolled in the Electrical Engineering Masters program. He has developed several biomedical signal acquisition systems and mission-critical industrial controllers. He works actively with start-ups in Brazil to foster embedded electronic device development and Internet of Things projects in the region.

Chirag Nagpal is currently in the junior year of Computer Engineering at the University of Pune, India. His research interests include machine learning and data mining. He is currently an intern at the Supercomputer Education and Research Centre at the Indian Institute of Science, Bangalore, where he works on problems involving the application of AI techniques to analyze data from social networks. He is a recipient of the Indian Academy of Sciences Research Fellowship and an Institution of Engineering and Technology scholarship.

Apart from his academic work, he enjoys hacking hardware and explores the developing area of the Internet of Things. Some of his work has been featured on Hackaday and Dangerous Prototypes. He has also been involved with the Texas Instruments Centre for Embedded Product Design at NSIT, Delhi, where he has trained undergraduates from across India on TI microcontrollers and development boards. He is also a licensed ham, with the call sign VU2CND. Complete details of his projects are available at http://www.chiragnagpal.com.

For this book, I would like to thank my parents and Prof. DV Gadre for constantly nudging me forward.

Dr. Philip Polstra (known to his friends as Dr. Phil) is an internationally recognized hardware hacker. His work has been presented at numerous conferences around the globe, including repeat performances at DEFCON, Blackhat, 44CON, Maker Faire, GrrCON, ForenSecure, and other top conferences. He is a well-known expert on USB forensics and has published several articles on this topic.

He has developed a penetration testing Linux distribution, known as The Deck, for the BeagleBone and BeagleBoard family of small computer boards. He has also developed a new way of doing penetration testing with multiple low-power devices, including an aerial hacking drone. This work is described in his book, *Hacking and Penetration Testing With Low Power Devices*, *Syngress*. He has also been a technical reviewer on several books including *BeagleBone for Secret Agents* and *BeagleBone Home Automation*, both by Packt Publishing.

He is an associate professor at Bloomsburg University, Pennsylvania (`http://bloomu.edu/digital_forensics`), where he teaches digital forensics, among other topics. In addition to teaching, he provides training and performs penetration tests on a consulting basis. When not working, he is known to fly, build aircrafts, and tinker with electronics. His latest happenings can be found on his blog at `http://philpolstra.com`. You can also follow him on Twitter at `@ppolstra`.

Glenn Vander Veer has been an embedded developer for over 15 years in various industries ranging from telecom to medical to electrical smart grids. He has developed code for a large variety of microprocessors and microcontrollers over this time. Lately, he has been developing code for electrical smart meters to aid in the collection of billing data and transmitting that data to the billing company.

www.PacktPub.com

Support files, eBooks, discount offers, and more

For support files and downloads related to your book, please visit www.PacktPub.com.

Did you know that Packt offers eBook versions of every book published, with PDF and ePub files available? You can upgrade to the eBook version at www.PacktPub.com and as a print book customer, you are entitled to a discount on the eBook copy. Get in touch with us at service@packtpub.com for more details.

At www.PacktPub.com, you can also read a collection of free technical articles, sign up for a range of free newsletters and receive exclusive discounts and offers on Packt books and eBooks.

https://www2.packtpub.com/books/subscription/packtlib

Do you need instant solutions to your IT questions? PacktLib is Packt's online digital book library. Here, you can search, access, and read Packt's entire library of books.

Why subscribe?

- Fully searchable across every book published by Packt
- Copy and paste, print, and bookmark content
- On demand and accessible via a web browser

Free access for Packt account holders

If you have an account with Packt at www.PacktPub.com, you can use this to access PacktLib today and view 9 entirely free books. Simply use your login credentials for immediate access.

This book is dedicated to the brave men and women from all corners of the civilized world who watch over us so that we may "sleep peacefully in our beds."

Thank you and stay safe.

Table of Contents

Preface

In this book, I have tried to include something for readers with various skill levels and interests. It contains hardware projects, software projects, and a combination of both. In all cases, I have tried to begin with a simple project and moved on to progressively harder and more complex projects. Depending on your skill level, some projects will take an hour or so to build and some will take longer.

Either way, I hope you will enjoy building the projects as much as I have enjoyed writing about them.

What this book covers

Chapter 1, Introduction – Our First Project, explains how to practice our soldering and de-soldering skills by building an infrared flash light and head lamp.

Chapter 2, Infrared Beacon, continues with our infrared light project — that is, building an invisible infrared flashing beacon.

Chapter 3, Motion Alarm, explains how to build a simple but effective intruder alarm with the always popular LM555 timer.

Chapter 4, Sound Card-based Oscilloscope, covers the beginning of the combined hardware and software projects. We will be building some hardware. It will allow you to use a USB sound card as a simple but useful oscilloscope.

Chapter 5, Calibrated RF Source, introduces you to the wonderful world of RF when you build a 50 MHz calibrated reference. Useful on its own, it will also be used in the next chapter.

Chapter 6, *RF Power Meter – Hardware*, shows us how to build ourselves a meter capable of measuring RF power at various frequencies. We will avoid most of the layout headaches by using a demo board provided by the manufacturer of the power detector.

Chapter 7, *RF Power Meter – Software*, explains how to build a BeagleBone Black-based software development system. Then, we will go on to write the software which will not only measure RF power but will also control an external RF attenuator.

Chapter 8, *Creating a ZigBee Network of Sensors*, covers how to build a wireless security system for your home or office based on the ZigBee RF module.

What you need for this book

What you need will depend on which projects you intend to build. The first four projects can be built with the basic hand tools described in *Chapter 1*, *Introduction – Our First Project*. The software referenced in *Chapter 4*, *Sound Card-based Oscilloscope*, was tested on an IBM PC-running Windows XP, so you do not need a powerhouse PC or laptop.

The next chapters will require access to some RF test equipment, such as a spectrum analyzer and possibly an RF signal generator.

The final chapters will require you to purchase a BeagleBone Black system. Some of the hardware is optional, but it is something I personally found useful. The reader should note that all of the software and hardware for this book was written on a PC running Windows XP.

Who this book is for

This book will hopefully have something of interest to a large variety of electronics enthusiasts, from hams to hackers.

I would say that, as long as you have at least intermediate programming and construction skills, you should have no problem completing the projects in this book. All the projects use through-hole parts to make assembly easier. All the files used to construct your own printed circuit boards, as well as all of the code, are available for download from the Packt Publishing website.

Conventions

In this book, you will find a number of styles of text that distinguish between different kinds of information. Here are some examples of these styles, and explanations of their meanings.

Code words in text, database table names, folder names, filenames, file extensions, pathnames, dummy URLs, user input, and Twitter handles are shown as follows: "What the preceding highlighted code does, is to first read AN0 using the `analogRead()` function."

A block of code is set as follows:

```
var app = require('http').createServer(handler);
var io = require('socket.io').listen(app);
var fs = require('fs');
var b = require('bonescript');

app.listen(8080);

console.log('Server running on: http://' + getIPAddress() + ':8080');
```

When we wish to draw your attention to a particular part of a code block, the relevant lines or items are set in bold:

```
var app = require('http').createServer(handler);
var io = require('socket.io').listen(app);
var fs = require('fs');
var b = require('bonescript');

app.listen(8080);

console.log('Server running on: http://' + getIPAddress() + ':8080');
```

Any command-line input or output is written as follows:

```
git clone https://github.com/ajaxorg/cloud9/
cd cloud9
npm install
chmod 777 .sessions
cd ~/cloud9/node_modules
```

New terms and **important words** are shown in bold. Words that you see on the screen, in menus or dialog boxes for example, appear in the text like this: "Once again, all the switches are handled in the same manner with the exception of the **Load** button."

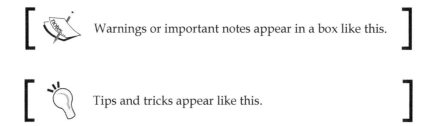

Warnings or important notes appear in a box like this.

Tips and tricks appear like this.

Reader feedback

Feedback from our readers is always welcome. Let us know what you think about this book—what you liked or disliked. Reader feedback is important for us as it helps us develop titles that you will really get the most out of.

To send us general feedback, simply e-mail feedback@packtpub.com, and mention the book's title in the subject of your message.

If there is a topic that you have expertise in and you are interested in either writing or contributing to a book, see our author guide at www.packtpub.com/authors.

Customer support

Now that you are the proud owner of a Packt book, we have a number of things to help you to get the most from your purchase.

Downloading the example code

You can download the example code files from your account at http://www.packtpub.com for all the Packt Publishing books you have purchased. If you purchased this book elsewhere, you can visit http://www.packtpub.com/support and register to have the files e-mailed directly to you.

Downloading the color images of this book

We also provide you with a PDF file that has color images of the screenshots/diagrams used in this book. The color images will help you better understand the changes in the output. You can download this file from: `https://www.packtpub.com/sites/default/files/downloads/4515OS_ColoredImages.pdf`.

Errata

Although we have taken every care to ensure the accuracy of our content, mistakes do happen. If you find a mistake in one of our books — maybe a mistake in the text or the code — we would be grateful if you could report this to us. By doing so, you can save other readers from frustration and help us improve subsequent versions of this book. If you find any errata, please report them by visiting `http://www.packtpub.com/submit-errata`, selecting your book, clicking on the **Errata Submission Form** link, and entering the details of your errata. Once your errata are verified, your submission will be accepted and the errata will be uploaded to our website or added to any list of existing errata under the Errata section of that title.

To view the previously submitted errata, go to `https://www.packtpub.com/books/content/support` and enter the name of the book in the search field. The required information will appear under the **Errata** section.

Piracy

Piracy of copyrighted material on the Internet is an ongoing problem across all media. At Packt, we take the protection of our copyright and licenses very seriously. If you come across any illegal copies of our works in any form on the Internet, please provide us with the location address or website name immediately so that we can pursue a remedy.

Please contact us at `copyright@packtpub.com` with a link to the suspected pirated material.

We appreciate your help in protecting our authors and our ability to bring you valuable content.

Questions

If you have a problem with any aspect of this book, you can contact us at `questions@packtpub.com`, and we will do our best to address the problem.

1
Introduction – Our First Project

In this book, we'll build several projects that will get progressively more challenging. You should be able to build the first few projects in an afternoon. A finished product may take you a while longer, depending on how professional you want it to look.

In order to build the first few projects in this book, you will need a few basic assembly skills and tools described as follows. The RF projects will require access to and knowledge of RF spectrum analyzers and/or RF power meters.

- You should be able to solder through hole parts
- You should be able to use a solder removal tool such as a solder sucker or solder wick
- You should be able to read the color codes of resistors
- You should be able to read voltage, resistance, and current with a multimeter

That's it folks! In order to build and use the first few projects, you don't have to know how to write a single line of code.

Basic tools

The basic tools that you'll need are as follows:

- Soldering iron or soldering station:

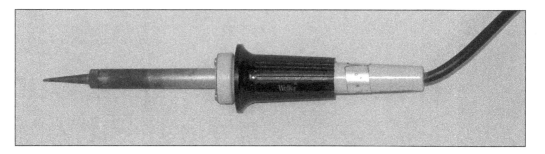

- Diagonal cutters:

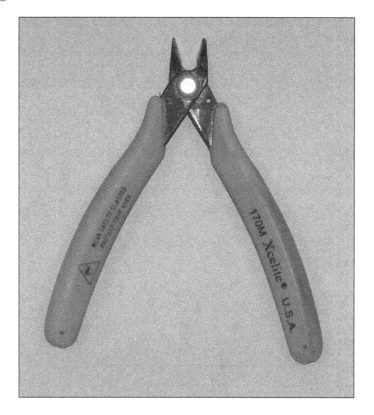

- Needle-nose pliers:

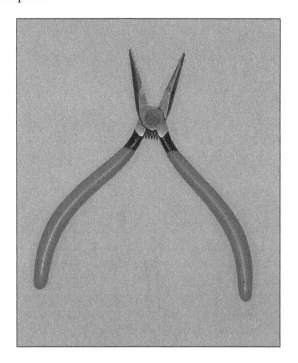

- Jewelers' screwdriver set:

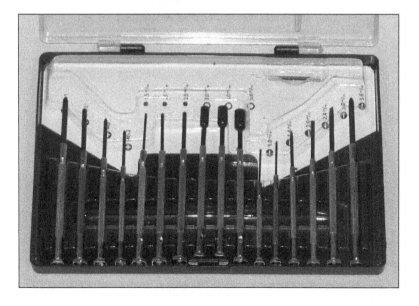

- Solder sucker or solder wick. Or both:

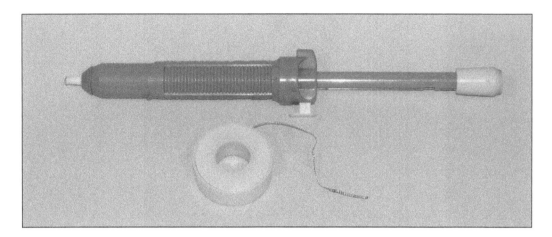

- A multimeter is capable of measuring volts, ohms, and milliamperes. The meter shown in the following image is a RadioShack/Micronta multimeter. If you have an opportunity to purchase one of these on eBay or at a Hamfest, I highly recommend you to do so. I have had this one for years. In addition to the usual meter functions, it also has a serial RS-232 output that has come in extremely handy over the years.

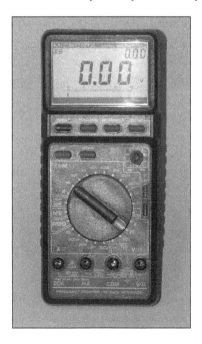

The more difficult projects will require access to either a BeagleBone Black or White, but we'll get to that later. Let's start our first project.

This project is the first in a series of projects for all you weekend warriors—paintball and Airsoft enthusiasts.

In this project, we will modify a standard LED flashlight that is available from eBay or your local variety store for a few dollars. What we are going to do is remove the visible light (white) LEDs and replace them with infrared LEDs. You might ask, Why would we want to do that? This is because many Airsoft and paintball rifles and pistols are equipped with a flashlight holder to allow the shooter to see his/her target in the dark. The downside to this is that your enemies can see you coming a mile away! So what we are going to do is make ourselves an infrared flashlight, which will allow us to see our target in the dark without it seeing us. Most inexpensive black-and-white video cameras can actually see well into the IR range, as can night-vision goggles. There are a number of inexpensive night-vision devices available from the sites that sell Airsoft gear, so I won't get into that here.

So, let's get started. This project basically involves desoldering and soldering. By the time we are done, you will be *really* good at both.

Dollar store flashlight

The preceding image is of the flashlight that I purchased at a local hardware store for about $2.00 Yours might not be exactly the same, but the principle of what we are about to do will work for just about any LED flashlight. All I would suggest is that you purchase one with as many LEDs as possible.

Flashlight – step 1

The first thing we have to do is to remove the plastic lens from the flashlight. We do this for two reasons: First of all, you will probably have to remove it in order to get to the LEDs. The second is that plastic tends to diffuse and attenuate infrared energy. I pierced the lens with a sharp object and popped it out, as you can see in the following image:

Lens removed

As you can see, the LED assembly is now loose in the case. If we remove the assembly and turn it over, we can see how the LEDs are connected to the battery pack. The positive or anode end of the LED is connected to the center terminal of the pack while the negative or cathode end of the LED is connected to the case.

Rear view of PCB

Flashlight – step 2

OK, time to warm up the old soldering iron and have at it. What you have to do is basically remove all the LEDs from the PCB without damaging it. Fortunately, these flashlights are cheap, so you can buy a couple of flashlights depending on your confidence level.

Once you have all the diodes LED's removed, it is time to replace them with the infrared ones. If you're lucky, the LEDs you are installing will have a flat side on the case. This is the cathode or negative side of the LED and it should be soldered to the negative side of the PCB.

Some LEDs have one short lead and one long one. The short one is supposed to be the cathode, but I would highly advise checking to be sure. The following image shows how to check your LED using the diode setting on your multimeter. In my case, it is indicated by an image of a diode.

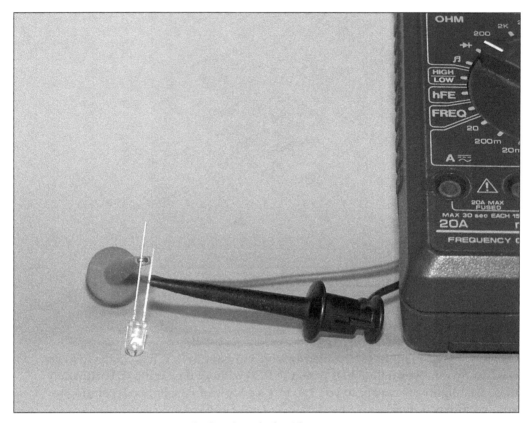

Finding the cathode with a meter

Checking an LED is like checking any other diode. With the negative lead of the meter connected to the cathode (short lead) and the positive lead connected to the anode (long lead), you should get the minimum reading of ohms.

Because LEDs do not behave like normal diodes, this test might not work. If you don't have a diode setting on your meter, just use the lowest ohm range. If you aren't sure, then the easiest thing to do is solder one LED and then temporarily reassemble the flashlight. If you cup your hands around the lens, you should see a faint red glow from the LED.

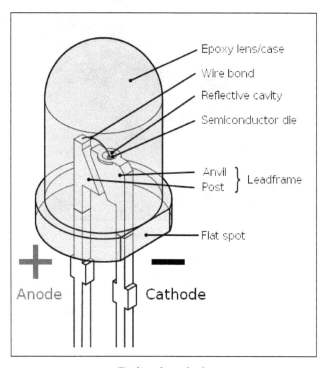

Finding the cathode

Now that you have the first one installed correctly, all you have to do is repeat the process for as many LEDs as you have. Once you have all of the LEDs installed, you should check the project by pointing the flashlight at a cheap black-and-white video camera. You should see the glow from the LEDs. One way to check if the camera will work is to point your TV remote at the camera and see if you can see the LED flash when you press a button on the remote.

Once you have tested your Special Ops flashlight, you should use 5 minute epoxy glue to hold the LED assembly in place.

Your finished flashlight will look much like the following image:

Finished flashlight

If you enjoyed this quick and hopefully easy project, there is another one similar to it in the following pages. Many of you have probably seen LED headlamps in camping and hardware stores. The following pages will show you how to modify one of these lamps in pretty much the same way as we did the flashlight.

If you're a nature lover rather than a weekend warrior, you might find this project useful because it will free your hands for your camera or binoculars.

Headlamp – step 1

The following image shows the headlamp that I used. Yours will probably look very much the same.

LED headlamp

We will be disassembling the headlamp in the same way as the flashlight. Simply turn the bezel until it comes out in your hands. Once again, we are talking about a $2.00 investment here so you might want to buy a couple, just in case you have to sacrifice a few in the name of science.

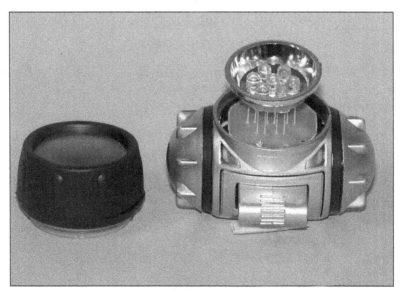

Lamp with lens removed

Once you remove the two small Philips screws, you can now remove the circuit board containing the LEDs. The headlamp I purchased came with a switch that allows you to have one, three, or all the LEDs on at the same time. I'm not sure how practical this is for our project, except possibly to conserve battery power.

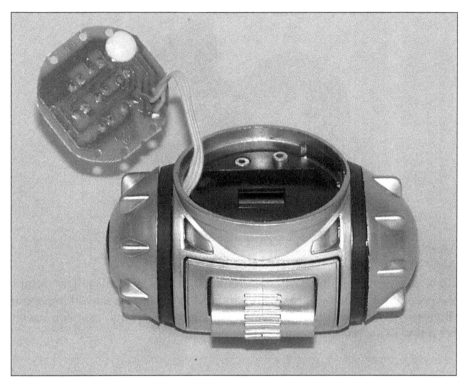

PCB removed for modification

Now that you have the lamp apart, it is time to fire up the soldering iron and solder removal tool and go to it. As in the previous project, simply replace the white LEDs with IR ones and off you go.

The cool thing about this lamp is the LED selector switch that allows you to turn on one, three, or all LEDs. If you are going to use it for nature observation, you could build yourself a lamp with three red LEDs and four white ones. That way, you can move about with the red LEDs and not ruin your night vision, unless you see something interesting.

Summary

This chapter contained two hopefully simple projects that allowed you to blow the dust off or enhance your soldering and desoldering skills. These skills will come in handy in the chapters to follow. We built two inexpensive but useful projects. All for about $5.00 per project, depending on how much the LEDs cost you.

Use your imagination. How about blue LEDs and wearing the lamp on your chest or head for Halloween?

In the next few chapters, your soldering and construction skills will be challenged even further, as we build a motion sensor out of a piece of a water pipe and a flashing infrared beacon. So read on.

2

Infrared Beacon

In this chapter, we will be building a flashing beacon. It's not just any old flashing light, though. This one will be *invisible*, unless you are wearing night vision goggles or viewing a black-and-white video camera.

What is a 555 timer and how does it work?

The integrated circuit that we will be using in this chapter has been around since the 1970s and is still going strong today. You can see the 555 timer circuit in the following diagram:

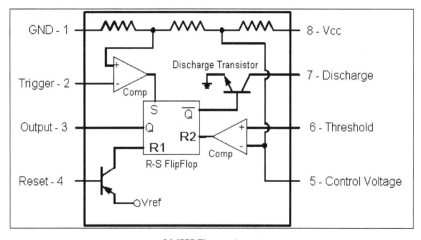

LM555 Timer pin out

The LM555 has two modes of operation. For this project, we will be using it in astable mode. The word **astable** means unstable, basically with respect to an oscillator. Now, in some cases we do not want the circuit to be astable, for example an amplifier. In this case, astable is a good mode.

The pins have various functions, depending on the mode in which we are using the device. The functions are as follows:

- Pins 8 and 1 are always power and ground respectively.
- Pin 3 is always the output pin that we will use to turn the LEDs ON and OFF.
- Pin 4 is the reset pin. Pulling this pin low (to ground) disables the oscillator in astable mode. Essentially the output goes low and stays low.
- Pin 5 is the control voltage pin. By applying a voltage to this pin, we can change the frequency in astable mode. By applying various signals to this pin, we can simulate UFO sounds from old science-fiction movies.
- Pin 6 is the threshold input to the on-chip comparator.
- Pin 7 is the discharge pin, used to discharge the timing capacitor.
- Pin 2 is the trigger input to a second on-chip comparator.

For those unfamiliar with how comparators work, a comparator *compares* the signals at its plus and minus inputs. If the positive input is higher than the negative input, then the output goes high. If it is lower than the negative input, then the output goes low.

For those of you who don't know what an R-S flip-flop is, it's the simplest of all memory devices. It's called a flip-flop because it has two complementary outputs that flip and flop back and forth, depending on the input. They are labeled **Q** and **Q/** because Q/ is always the opposite of Q. Hence the name flip-flop.

When a high level (logic 1) is applied to the Set (S) input, the Q output goes high and the Q/ output goes low. When a high level is applied to the Reset (R) input, the Q output goes low and the Q/ output goes high.

The R-S flip-flop truth table as follows:

Set	Reset	Q	Q/
1	X	High	Low
X	1	Low	High

The schematic for the circuit is shown in the following diagram:

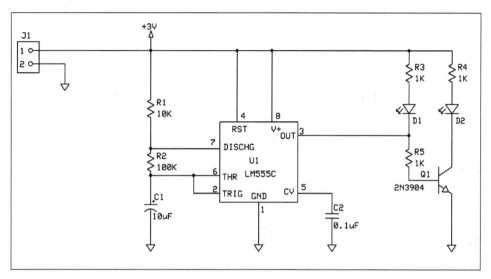

Flasher schematic

In our flasher circuit, timing resistors **R1** and **R2** and capacitor **C1** control the flash rate. The two comparator inputs, pin 2 and pin 6, are connected together.

When the output is high, the internal discharge transistor is turned off and the voltage across **C1** increases until it reaches two-thirds of the Vcc, at which time the comparator output on the trigger terminal becomes high. This resets the flip-flop, which causes the timer output to go low.

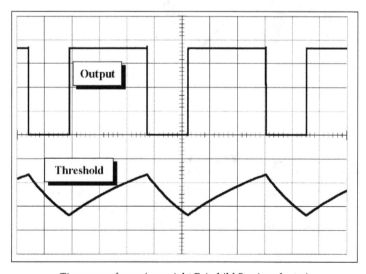

Timer waveforms (copyright Fairchild Semiconductor)

This turns on the discharging transistor and the capacitor is discharged through the transistor and R2. Once the voltage across C1 falls below two-thirds of the Vcc, the comparator output on the trigger terminal goes high, as does the timer output. The discharging transistor turns off and the voltage across C1 begins to rise again.

The time that the output is high is approximately: *Th = 0.693(R1+R2)xC1.*

The time that the output is low is approximately: *Tl = 0.693xR2xC1.*

Since R2 has the most effect on the timing, if we make it a lot larger that R1, we will get a square wave output. That is, the on and off times will be the same.

You can vary the on and off times by playing with various combinations of R1 and R2. For example, you can have a one long or one short flash rate.

Our 555 timer circuit

In our circuit, when the output is high, it turns on **Q1**, which turns on **D2**. When the output is low, it sinks the current, which turns off **Q1** and turns on **D1**.

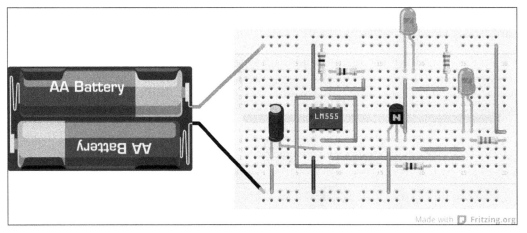

Breadboard layout

Assembling our flasher PCB

The following figure is of the PCB design that is available for download from the Packt site (`https://www.packtpub.com`). This PCB is designed to fit inside a tube with a one-inch internal diameter. You will see the reason for this in a moment.

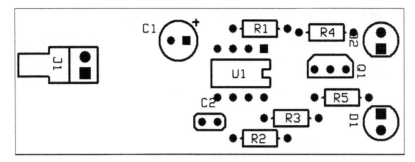

Flasher PCB

Assembly of the PCB is straightforward. I would suggest that you solder in the lowest parts first and work your way up to the higher ones. This way, the parts won't fall out when you turn the board over to solder them in. In this case, install the resistors first; then the LM555; and finally the capacitor and the transistor. You can also see how the LEDs have been bent at 90 degrees.

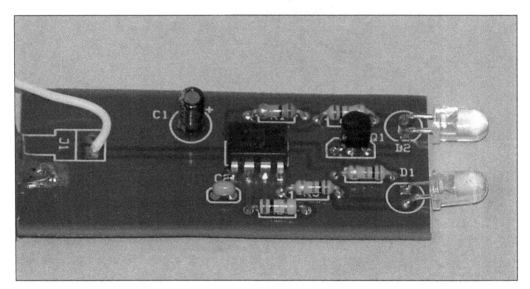

Finished PCB

Building and assembling the case

The flasher is designed to fit inside a piece of one-inch PVC plumbing pipe.
The LEDs are bent at 90 degrees to the PCB and inserted through holes drilled in
a one-inch PVC pipe cap. A #7 drill bit is used to drill the holes for the LEDs. The
cap is then turned over and a quarter of an inch drill bit is used to enlarge the rear of
the holes so that the LEDs will penetrate further into the cap. You should drill about
half way through the cap with the quarter of an inch drill. A drill press with a depth
gauge will come in handy here.

Cap showing drilled holes

One end of the case is sealed shut with another one-inch end cap, which is glued into
place. The cap with the holes is not, so that you can change the batteries. Also, I did
not include an on-off switch in the design for the sake of simplicity, and to make the
case as moisture-proof as I could.

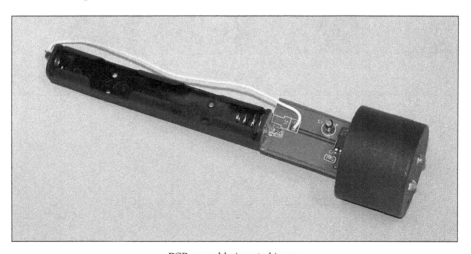

PCB assembly inserted in cap

The preceding image shows the completed PCB assembly installed in the end cap. These are prototype PCBs, so the ones you download from Packt might look slightly different. Basically, what I did is to remove some of the solder mask from the top ground plane and solder the negative terminal of the battery case to the top copper layer of the board. The white wire is the positive connection. This is where you would put the on-off switch if you wanted to.

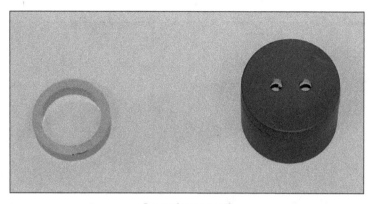

Cap and spacer ring

The cap assembly consists of the drilled cap and a spacer ring about one-quarter of an inch thick, cut from the one-inch diameter pipe. The spacer is used to allow you to glue the PCB to the cap and to make it easier to remove the cap, in order to insert and remove the batteries.

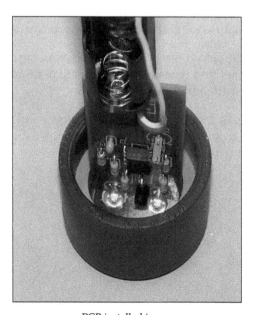

PCB installed in cap

The preceding image is a close-up of the board installed in the cap. You will notice that the ring comes up the side of the PCB by about one-quarter of an inch. This is to allow you to run a bead of epoxy along either side of the board. If you are reading the digital version of this book, you should be able to zoom in and see that the LEDs are recessed into the cap. That's why we drilled quarter of an inch holes.

The following image is of the finished project:

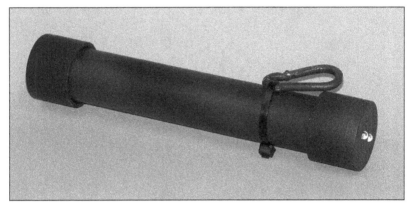

Finished flasher in case

Going further

There are a number of modifications that you can make to this project. For example, it is possible to fit a miniature toggle switch into the top cap, alongside the LEDs.

It is also possible to fit the circuit into a smaller case by using a 3 V lithium coin cell. I chose PVC pipe because I find it easier to machine than other materials and through hole components for first-time builders.

The following image is of a much smaller device. Patents and non-disclosure agreements prevent me from showing you anymore; however it is an example of what can be achieved with a little ingenuity.

Small IR flasher

Summary

In this chapter, we learned how LM555 timers work in astable mode. We built and tested a simple but useful circuit to demonstrate this.

In the next chapter, we will be building another "secret agent" device: a covert motion detector.

3
Motion Alarm

In this chapter, we will once again use the venerable 555 timer. This time, we will be building a hidden motion sensor. By hanging several of these little gems around your camp, you can keep your fellow paintball opponents from sneaking up on you!

How a 555 timer works in monostable mode

In this case, we will be using the LM555 in a modified monostable mode. Actually, we will not be using either comparator in this design. The only timing part of LM555 that we will be using is the internal R-S flip-flop (see the *What is a 555 timer and how does it work?* section in *Chapter 2, Infrared Beacon*) in the following diagram:

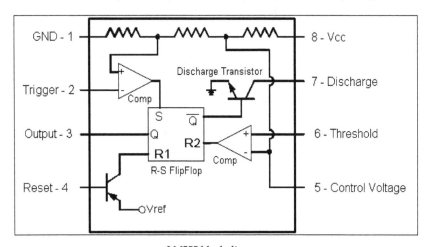

LM555 block diagram

How our alarm works

Here's how our motion-triggered alarm works. It uses a mercury switch like the kind used in car alarms. The mercury switch is connected to terminals W3 and W4. When the switch is disturbed, it grounds the trigger input, which causes the output to go high. This turns the buzzer on.

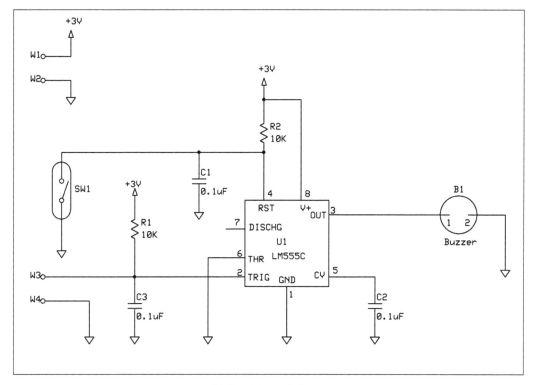

Motion-triggered alarm

The alarm will continue to sound until the internal flip-flop is reset. We do it by placing a magnet within the range of the reed switch. Closing the switch grounds the reset pin and turns off the buzzer by making the output go low. It's that simple. A suggested breadboard layout is shown in the following image:

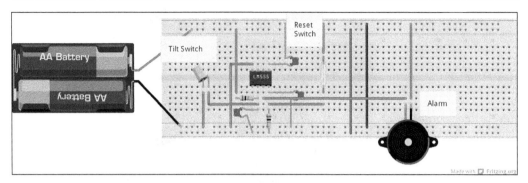

Motion alarm breadboard

Assembling the case

The first thing we have to do is make ourselves the end cap that will have the PCB fastened to it. This cap has a half-inch hole drilled in it for the buzzer. I made the grill from a beer coaster I got from the local dollar store, but any coarse fabric will do.

We will be using a PVC water pipe (with a diameter of three-quarters of an inch) as our case. You should cut yourself one piece about eight inches long. You will also need a piece of pipe about one-quarter of an inch long as a retaining ring. I used a coin as a template to cut a grill with a diameter of three-quarters of an inch. These are shown in the following image:

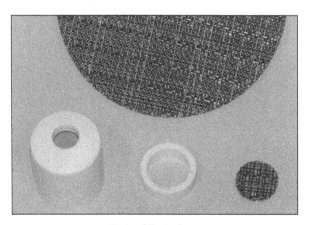

Parts of the end cap

From left to right, we see the end cap, the retaining ring, and the grill cut from the coaster, which is in the background.

The next step is to fasten the grill in place with a couple of beads of epoxy.

End cap with grill installed

Once the epoxy has set, it's time to insert the small one-quarter of an inch ring that we made earlier. This ring will hold the PCB and buzzer in place. Carefully push the ring into the cap, making sure that it goes in straight. You might have to use the 8-inch tube that you cut for the body, to push it to the last bit. The finished cap should look like the following image:

Cap with spacer installed

Assembling the PCB

Assembling the actual PCB is straightforward. Just solder all of the parts except the mercury tilt switch into their locations marked on the silkscreen.

The reed switch is marked **SW1**. The mercury tilt switch is connected to pads **W3** and **W4**. The terminal marker plus or + of the buzzer (**B1**) goes to the square pad of the component (pin 1).

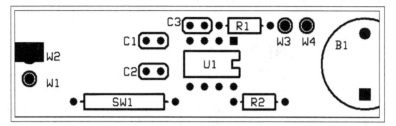

Alarm PCB silkscreen

The mercury tilt switch is installed as shown in the following image. This board is a prototype. Your PCB from the Packt site will have the pads relocated so that you won't need the wire. The switch itself mounts in the same direction as the one in the image, and should be held in place with epoxy or hot glue. The switches I got were designed for car alarms and were extremely sensitive.

For those of you who don't know how a mercury switch works:

Mercury is the only metal that is liquid at room temperature. As with any liquid, surface tension causes it to form a sphere. This shape makes it roll around easily. If we use a small amount of mercury, we get a small drop with very little mass that rolls around very easily.

A mercury tilt switch is just a glass tube with two contacts at each end. When the mercury bead rolls to the contact end, it completes the circuit (because it's a metal.) That's all there is to it.

PCB and battery assembly

I removed some of the solder mask from the top copper ground plane and soldered the negative lead of the battery holder to the ground plane. The positive lead of the holder is connected to W1 on the circuit board. Fasten the battery holder to the PCB with epoxy or hot glue and when it dries, you are ready for the next step.

Testing the alarm

Now that we have a finished PCB assembly, it's time to test it. For these tests, you will need a relatively strong magnet and two AA batteries. A magnet salvaged from a car speaker or a rare earth one from a local hobby supplier should work fine. It needs to be fairly strong, so a fridge magnet probably won't work. You are about to find out if your magnet must be big enough by following the given steps:

1. Make sure you have the magnet handy and then insert the AA batteries into the holder. There is no polarity protection, so be careful; the buzzer should be squealing at this point.

2. Stand the PCB assembly on end, with the buzzer at the bottom.

3. Wave the magnet by the reed switch and the howling will stop. Your alarm is now armed.

4. If you tilt the PCB assembly left or right, the squeal should start again.

This test should give you some idea of just how sensitive this alarm is!

This is an extremely simple circuit, so there is very little to troubleshoot. If things don't work as I said, I suggest that you check the polarity of the buzzer or the batteries.

[

If you want to make your circuit more foolproof, you can insert a small switching diode in series with the white wire. The cathode (band) of the diode connects to W1. This will give you reverse polarity protection in case someone installs the batteries backwards.
]

Now that we have a working PCB assembly, we can continue with the assembly. What we now have to do is install the assembly in the case we assembled earlier. The following image shows how the board should be attached to the cap:

PCB assembly in the cap

You might find it easier if you cut a second one-quarter of an inch ring and insert it in the cap. That way, the ring will come further up the PCB. It really depends on your mechanical assembly skills. Basically, the idea is to add several layers of epoxy until the board is held firmly in place. Allow the epoxy to dry between layers and be careful not to plug the hole in the buzzer.

What we are building here is a removable assembly that is attached to one of the caps. This cap will *not* be glued to the tube, so that we can change the batteries and get to the circuit if need be.

The other cap has a ring attached to it that allows you to hang the alarm from a tree limb or whatever you have. Just like a Christmas decoration!

End cap with a ring

To deploy these little gems, I would suggest that you insert the batteries and then hold the magnet near the alarm when you hang it; either that or use a piece of tape to cover the speaker hole, and then gently remove it. I'll leave the details of the installation up to you.

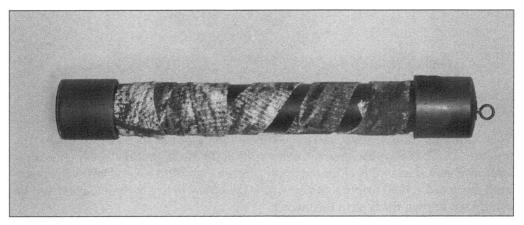

Assembled alarm

The preceding image shows a finished alarm. If the alarm is not triggered, the batteries should last for quite a while, so I suppose you could use them as a cheap perimeter alarm at home or in a dormitory.

Going further

As it stands, the LM555 timer turns on the buzzer, but that does not mean that it cannot control a transistor or a relay. With an RF remote control, you can turn on floodlights or a bigger siren. I will leave the rest up to the reader.

Summary

In this chapter, we learned how to use the LM555 in a completely different way. We used only a part of the chip to build ourselves a very simple but effective motion-sensing alarm. What we have here is basically a repackaged car alarm.

In the next chapter, we will be building an oscilloscope interface for your USB sound card.

4
Sound Card-based Oscilloscope

The title of this chapter is *Sound Card-based Oscilloscope*; however, I am going to start by highly recommending an external USB sound card such as this one from Creative Labs:

Model SB-1090 from Creative Labs (Image courtesy: eBay)

This model from Creative Labs (`http://us.creative.com/p/sound-blaster/sound-blaster-x-fi-surround-5-1-pro`) has very impressive specifications. Whether you decide to go with an internal or external card, you will want one with similar specifications:

- **Playback**: Up to 24 bits/96 kHz 5.1

- **Signal-to-Noise ratio (SNR)**: Less than 100 dB

- **Total harmonic distortion + noise at 1 kHz**: 0.01 percent

- **Recording**: Up to 24 bits/96 kHz

Note the 96 KHz sample rate. This gives us a maximum measurable frequency of 48 KHz at an SNR of less than 100 dB. You most likely won't get these specifications from an internal card. Get the highest sample rate and highest SNR your budget will allow.

The output section

We will start with a description of the output section of the project. I know that oscilloscopes don't have outputs but sound cards do, and a number of software vendors have used this to their advantage. We will cover the details of that subject later. In the meantime, let's look at the output circuit.

48 KHz is only a theoretical maximum. The actual practical maximum frequency you are able to measure might be much lower. For information on the Nyquist–Shannon sampling theorem, refer to `http://en.wikipedia.org/wiki/Nyquist%E2%80%93Shannon_sampling_theorem`.

The following circuit diagram shows the output buffer circuit. This circuit is used to isolate the soundcard from the circuit being tested.

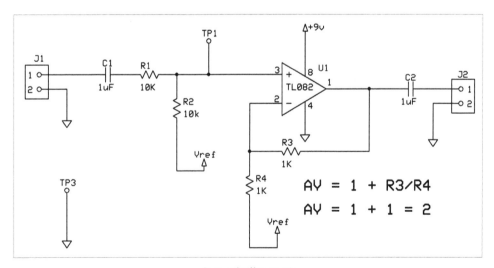

Output buffer circuit

The circuit may look odd to some of you, but there is method in our madness. Many operational amplifiers have extremely high bandwidth at unity gain. This causes them to oscillate. So the easiest way to avoid this is to add some negative feedback.

In this case, resistors R1 and R2 form a 2-to-1 voltage divider. The gain of the amplifier is set to 2 by resistors R3 and R4 for a net gain of zero from the input to the output. The reason we must have unity gain is that the software assumes that you are connecting your sound card directly to the circuit you are testing. We have no way of telling the software that there is a gain stage in between.

The output impedance (resistance) of the TL082 operational amplifier is quite low; so, if you want your project to have a 600-ohm output impedance, you can just add a 620-ohm resistor (or two 300-ohm resistors) in series with C2.

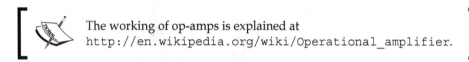

The working of op-amps is explained at
http://en.wikipedia.org/wiki/Operational_amplifier.

The PCB for the output section has both channels on the same board. I chose to make this system modular for several reasons. The good folks at ExpressPCB have a deal on boards of a certain maximum size. Making the system modular helps keep the price down for you and allows you to build only the modules you want. The input circuitry is more complex, so one board is used for the input and another for the output.

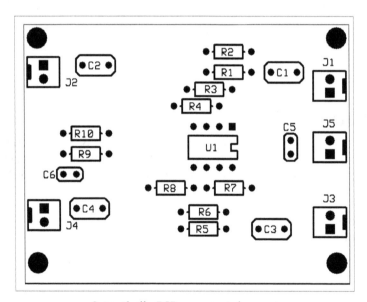

Output buffer PCB component placement

The PCB file that comes with this book has two of these boards on the same panel, so you can use them for additional channels or whatever you like.

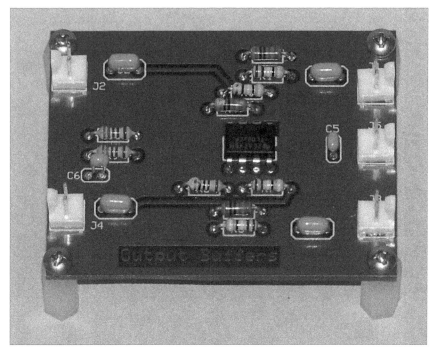

Assembled output buffer PCB

The input section

The input amplifiers are a bit more complex. The input is AC coupled with U1, which has very high input impedance. AC coupling means that the DC component (offset) from any input signal will be removed. The gain of the first amplifier is *R2/R1*, or 1 in this case.

Oscilloscope probes come in times-one or times-ten varieties. In fact, some have an **X1/X10** selector switch, as shown in the following diagram:

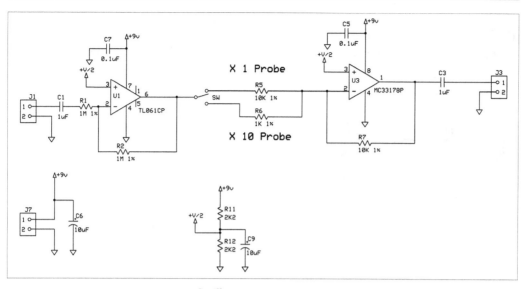

Oscilloscope input amplifiers

A times-ten oscilloscope probe multiplies the input impedance of your scope by a factor of 10. So your 1-megohm input scope now has an input impedance of 10 megohm. However, since nothing in life is free, it also reduces the signal level by a factor of 10. So, if it is not corrected, a 1-volt signal will appear as a 0.1-volt signal on your scope. Many modern scopes do this automatically. Ours has a switch instead.

The reason modern oscilloscopes have this option is that a 10-megaohm probe will tend to load a high impedance less than a 1 megohm probe and, therefore, affects the measurement less.

For information on input impedance, refer to http://en.wikipedia.org/wiki/Input_impedance.

When the switch is in the **X10** position, meaning a times-ten probe is being used, the output is multiplied by 10 to compensate for the probe's loss. When it is in the **X1** position, the output is multiplied by 1.

The input amplifier is an inverting amplifier and the output amplifier U3 is also an inverting amplifier, so the net result is no signal inversion. I chose this configuration because the gain of an inverting amplifier is easier to obtain with standard resistor values.

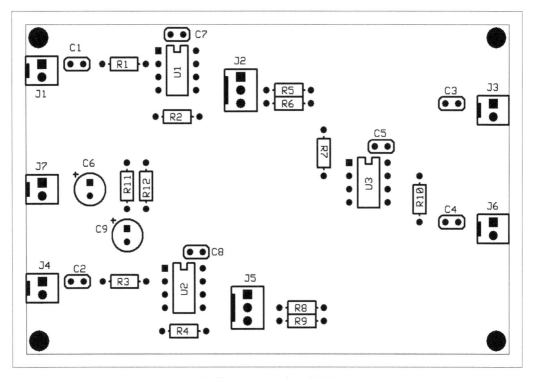

Oscilloscope input board PCB

Resistors R11 and R12 provide a mid-range voltage reference of about 4.5 volts. This means that the AC signal will have a + 4.5 V offset. This is done so that the signal can have a +/- 4.5-volt swing. This is also why the output capacitor, C3, is required — so that the signal will once again swing +/- 4.5 volts about the ground voltage.

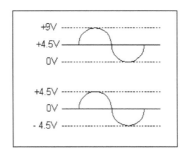

Signal offset

In the preceding diagram, the top trace shows the internal voltage swing and the bottom trace shows the output from the PCB with the DC offset removed.

The PCB assembly

Now that we have a better understanding of how the circuitry works, it is time to heat up our soldering irons and get to work on assembling the actual hardware.

Here we have the two PCBs ready for final assembly. Notice that the 5/8" standoffs on the output board, on the left, have 4 to 40 threaded studs on the bottom. You will see why in the following image:

Assembled PCBs

In the preceding image, we see the two boards mounted one on top of the other. I had to do this in order to make them fit in the case I chose. I also had to use 90-degree mounted connectors. Depending on the case you choose, you may or may not have to do this.

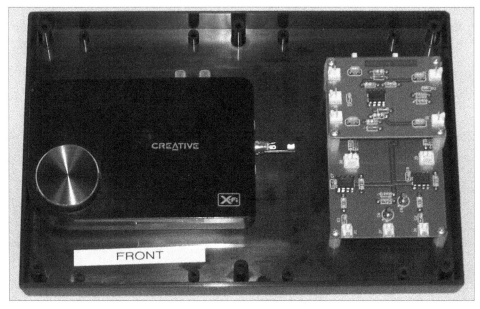

PCBs and SoundBlaster in a case

Here we see the SoundBlaster and PCBs installed in the case and ready for the wiring harnesses. I would recommend that the SoundBlaster be held in place with Velcro for easy removal. Once you have it installed, you are ready for the final assembly.

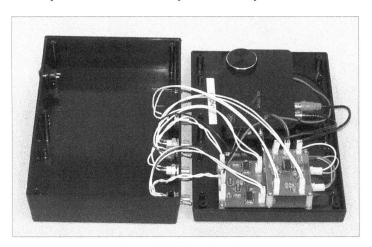

Case with PCBs and wiring

In the preceding image, we see the case almost completed. I left the USB and power connections out of the image for clarity. Notice that the wires are long enough for the lid to lie flat next to the base.

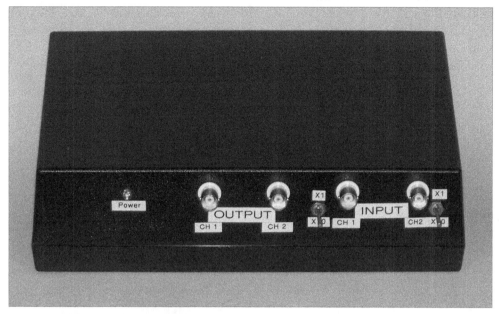

Finished project

The previous image shows my finished prototype. If you use the parts in the bill of materials, your prototype will look very much like this.

Software

Now that we have the hardware built, it is time to put it to use. A simple Google search turned up several free and commercial packages. I have included the free ones for download from the Packt Publishing site.

Sound card oscilloscope program

One of the more useful packages I found was called **sound card oscilloscope,** which is available for download from `http://www.zeitnitz.de/Christian/scope`. The following screenshot was taken from this package when it was in use. The only downside of this package is that it does not use the full 96 KHz sample rate of the sound card I am using.

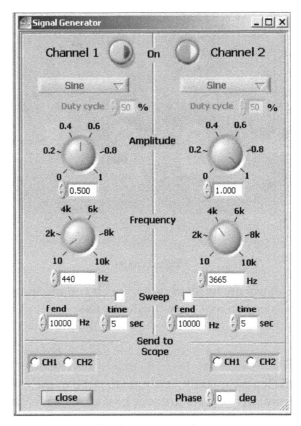

Signal generator window

The preceding screenshot shows the signal generator window, which is detachable from the main scope window — a handy feature in my opinion.

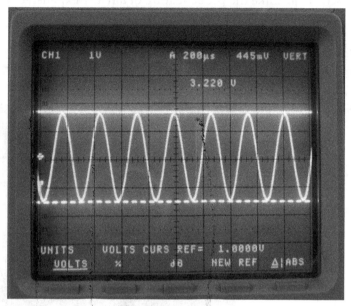

An oscilloscope

The preceding image shows the resulting output from our black box. Unfortunately, we only have a 48 KHz sample rate, so the highest frequency we can generate is about 24 KHz. Also, the software can only use 16 of the 24 bits, so the sine wave is not as clean as it would have been if we had been able to use all 24 bits.

You will recall that the interface has a times-ten and a times-one probe selector switch. The following screenshots are of the main oscilloscope screen, with the switch set to the X1 probe and then the X10 probe.

Notice that, when the switch is set to the X10 probe, the signal is larger by a factor of ten than when the switch is in the times one position. This is to compensate for the loss caused by the probe when it is set to X10.

With your probe set to X10 and the switch in the X10 position, you now have an oscilloscope with an input impedance of 10 megohms that displays the correct reading on the screen, just like the probe feature of a really expensive commercial oscilloscope. Isn't that cool?

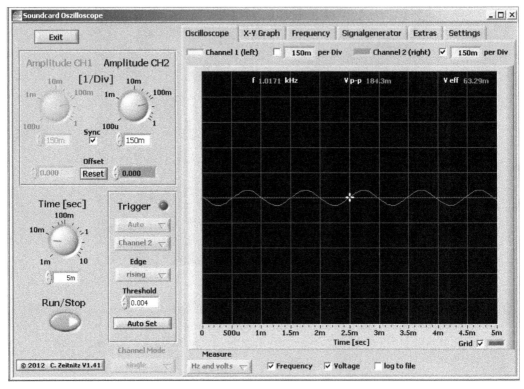

Switch set to X1

Here we see the results of the switch set to X1 and an actual input voltage of about 184 mV peak to peak.

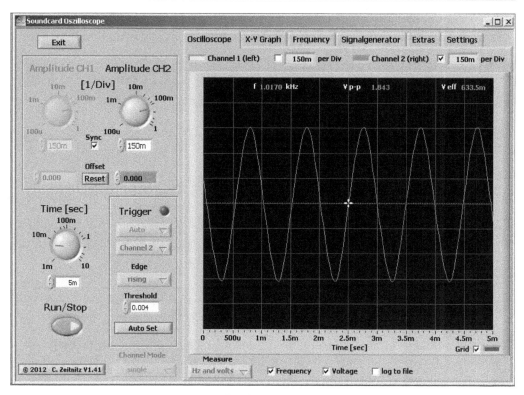

Switch set to X10

Notice in the preceding screenshot that the signal is now 10 times greater than the previous signal, again to compensate for the loss of a times-ten probe.

The Zelscope software

Another interesting commercial package is the one available from Zelscope, and a 14-day trial version can be downloaded from their website at http://www.zelscope.com/.

This is a commercial package but it has some very useful features. The main window looks very much like (and has many of the same features as) an actual oscilloscope.

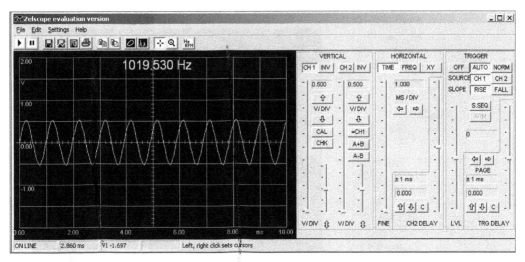

The Zelscope oscilloscope window

The software has a number of useful modes including an audio spectrum analyzer. The following screenshot shows the same 1 KHz sine wave shown in the preceding screenshot in the spectrum window:

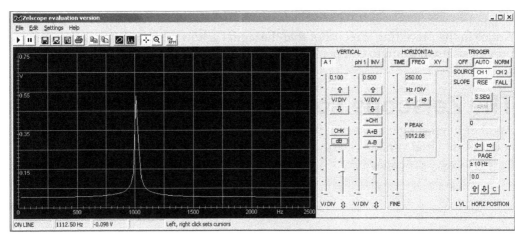

The Zelscope spectrum window

Visual Analyzer

In my search for interesting software, I found a very interesting and useful package called Visual Analyzer. It is written by an Italian gentleman named Alfredo Accattatis (http://www.sillanumsoft.org/intoduction.htm).

The software can be downloaded from http://www.sillanumsoft.org/download.htm.

The software is "donate ware," which means that it is free for you to use but you get additional support if you make a donation via PayPal. It is well worth a generous donation in my opinion. The following screenshot is of the main window.

One annoying thing that I discovered about the site is that the pages are not linked together very well, so I am including useful links wherever I can.

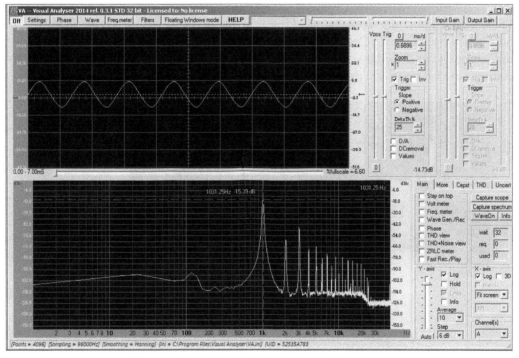

Main screen of the Visual Analyzer

The main window displays two functions simultaneously. The top half is the oscilloscope view and the bottom half is the audio spectrum analyzer view.

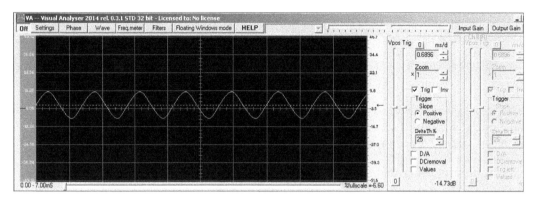

Oscilloscope window

The oscilloscope window contains all the controls that you would expect to find on a bench oscilloscope. One thing I found handy is that both the zoom and the milliseconds per division were variable. For example, you can set the zoom to X10 to compensate for your X10 probe, and you can also compensate for losses in a test jig.

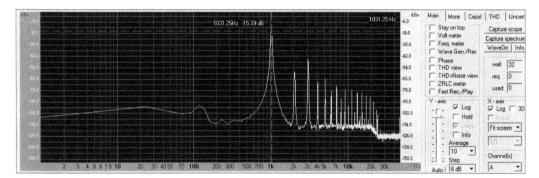

Spectrum analyzer screen

Next to the spectrum analyzer window is the number of tabs. Only the frame labeled *y*-axis controls the spectrum analyzer. I discovered that some of the features work and some don't in this version of the software—something that I am sure will be fixed in future revisions of the software.

A more complete description of all the features is available at http://www.sillanumsoft.org/prod01.htm.

The ZRLC meter

One feature that I did find fascinating though was a built-in ZRLC meter. This meter measures impedance (Z), resistance (R), inductance (L), and capacitance (C).

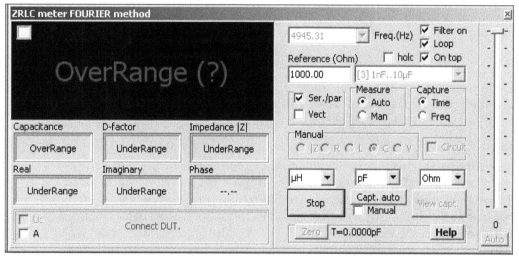

ZRLC meter

The software's author describes a simple test fixture for use with this feature. After some digging, I managed to find a schematic for the circuit he was referring to. The schematic is shown in the following diagram.

ZRLC hardware

The following diagram shows an improved version of the hardware we will be using:

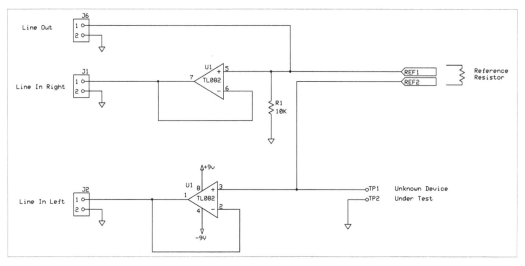

Improved schematic

How the circuit is used to measure ZRLC is described at
http://www.sillanumsoft.org/ZRLC.htm.

However, I have made some of the improvements that he suggested and corrected an error on the schematic he included. My circuit works mostly the same as he described. The TL082 operational amplifier has a **FET** (short for **Field Effect Transistor**) input, which gives it a very high input impedance. This is important because as the author says, the input and output impedances of the sound card adversely affect the readings.

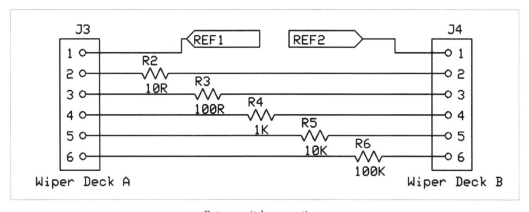

Rotary switch connections

A rotary switch allows you to select various ranges for the jig. You will notice that the reference (ohm) control is a textbox, so you can either use 1% resistors or measure your resistance with a good digital meter and, theoretically, get even better accuracy.

Making ZRLC measurements

The test procedure on the web page is a bit hard to follow, so I have included my own version here. I have not had an opportunity to build and test the fixture, so the following steps are based on my understanding of how the jig is supposed to work:

1. Make sure that the **Device Under Test** (**DUT**) is not connected to the jig.

2. Choose a proper range, starting from the lowest if you are not sure of the value of the device.

3. Click on the **Measure** button and the ZRLC meter will begin its calibration cycle.

4. Wait until the **Connect DUT** message appears.

5. You should see the **Over Range** message on the display. This is because you haven't connected the unknown device yet.

6. Now you are ready to make the actual measurement. Connect the DUT and click on **Measure** again.

   At this point, one of these four will happen:

   ° The measurement is correct and you simply read the value from the display.

   ° The measurement is OK but a (u) symbol appears. In this case, switch to the next higher range and start again from the calibration step.

   ° The measurement is OK but an (o) symbol appears. In this case, switch to the next lower range and start again from the calibration step.

   ° If the **Under Range** or **Over Range** message does not go away, then the value is much higher than your selected range, and you should start over with a much higher or lower range setting.

According to the website, the jig and the software are capable of measuring the following:

* Resistance

* Impedance (real and imaginary part):
  http://en.wikipedia.org/wiki/Electrical_impedance

* Capacitance

- Inductance
- Input impedance of amplifiers, transformers, and so on
- All the previous parameters at different frequencies, and with automatic sweep, in-time, and frequency domains

As I said, time and funding did not permit me to actually build and test the jig described on the aforementioned web page, so I cannot testify to its accuracy.

Using the sweep generator to measure frequency response

There is another useful function of this software that I was able to test: the sweep generator. This feature will allow you to measure the frequency response of a filter or amplifier.

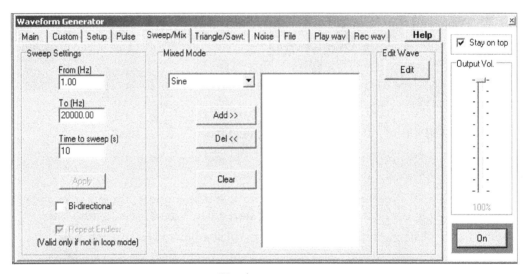

Waveform generator

Peak detector hardware

There is another useful piece of hardware you can build that will work with any sweep generator and almost any software. It is called an **active peak detector**. It takes in an audio signal and outputs a DC level that is approximately equivalent to the peak-to-peak voltage of the input signal. This is the type of circuit that drives the VU meter on your stereo equipment.

The reason it is called an active detector is that it uses operational amplifiers to buffer the input signal so that the circuit does not load the source.

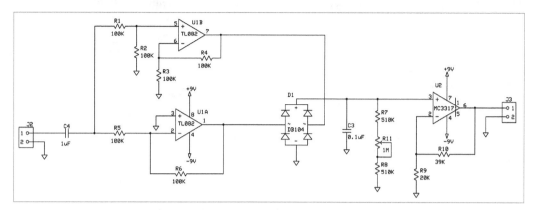

Peak detector

In the preceding diagram, amplifier U1 is used to create an active transformer that feeds the AC inputs of the full wave rectifier. The amplifier, U1A, is an inverting amplifier with a gain of 1 and U1B is a non-inverting amplifier with a gain of 2. That is why we divide the input signal to U1B by 2. The result is two AC sine waves 180 degrees out of phase, just like the secondary of a transformer.

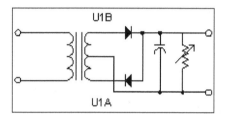

Simplified peak detector

Capacitor C3 and resistors R7, R8, and R11 form a simple filter or envelope detector. The variable resistor, R11, sets the time constant (a measure of how fast the capacitor C3 discharges) of the circuit. Amplifier U2 is just a buffer and it can have any gain you want.

Example test setup

The following diagram shows a simple test setup used to measure the audio frequency response of a device:

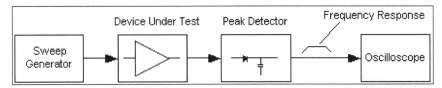

Test setup to measure the audio frequency response

In the preceding diagram, the device is shown as an amplifier, but it could just as easily be an active or passive filter, for example a speaker cross-over network.

In any case, the basic technique is the same. What you do is set up so that the time for one sweep and the time base of the oscilloscope are the same. This means one sweep covers the full width of the oscilloscope screen.

You hook the test as shown and sweep away. You may have to play with the sweep times and the oscilloscope time base to get an accurate display. Unless you can find a generator program with a logarithmic sweep, the entire response might not fit on one screen. You may have to sweep the test circuit in bands such as high, low, and mid-range in the case of a simple speaker crossover network mentioned earlier.

Summary

We covered a great deal of content in this chapter. We built and tested a frontend for a software oscilloscope and an audio signal generator. We also looked at several useful software packages available for free in some cases. We also saw some additional hardware that allows us to use the sound card to measure impedance, inductance, resistance, and capacitance. In addition we also learned how to build an active rectifier / envelope detector.

You will find the ZRLC and envelope detector printed circuits on the same panel for manufacturing. That way, if you order your boards from EasyPCB, you will get three of each. In the next chapter, we will be venturing into the mystical world of RF design. In spite of what some HAMs may tell you, there is nothing magical happening here (at least at these frequencies)—just a few simple rules that, if followed, will lead to success.

5

Calibrated RF Source

We will be building the first of two RF projects in this chapter. In order to calibrate the power meter in the next chapter, we first have to build ourselves a calibrated RF reference. I am going to assume that you know how to make measurements using a spectrum analyzer. If you're building an RF project for the first time, I'm sure a local ham radio club will be happy to help you out. They'll probably want to build one themselves.

What we are going to do is use a factory-programmed crystal oscillator (Digi-Key) and another crystal of the same frequency as a filter. The manufacturer's part number for the oscillator is MXO45HS-3C-50M0000. The other parts are listed in the bill of materials, which is available for download along with the PCB and schematic files.

A crystal is simply a piece of quartz that has been cut and tuned to a specific frequency. A crystal oscillator contains a tuned crystal and other circuitry to generate a stable frequency that is not subject to external factors such as circuit loading.

All RF oscillators have one thing in common: a tuned section that determines the frequency of the oscillator. Many oscillators use an inductor and a capacitor as the tuned section. In this case, either the inductor, the capacitor, both can be adjustable. We won't be using this type of oscillator in this design, so I won't waste any more of your time, except to say that crystal oscillators tend to be more stable and more accurate.

How good a filter is is based on what is called the Q of the filter. The higher the Q, the better the filter. The Q of an L-C (inductor/capacitor) filter is reduced by the resistance of the wire in the inductor and, in some cases, the inductance of the capacitor. I know that the last statement sounds strange to you, but some capacitors are constructed by wrapping a foil and an insulator into a tube-like shape. Believe it or not, at higher frequencies this gives the capacitor an inductive property. By using very good (expensive) components and some other tricks, we could build a fairly good filter. But, for the price of one inductor, we can buy ourselves a crystal that will act as a very high-Q filter.

What is Q or Q-Factor?

For more information, go to `http://www.radio-electronics.com/`
`info/formulae/q-quality-factor/basics-tutorial.php`.

You're probably wondering just why you need a filter. The reason is that the oscillator we will be using is designed to supply the system clock for a microprocessor or CPU. What this means is that the output resembles a square wave. For those of us that didn't take digital signal processing in college, a square wave is creating by summing, or mixing together, odd harmonics of the fundamental (center) frequency. So, in order to get just the fundamental frequency, we pass the square wave through a high-Q filter. This filter removes the harmonics and leaves just the fundamental frequency.

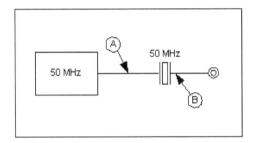

Diagram of a crystal filter

The following diagram shows the signal before (**A**) and after (**B**) the crystal filter. You should see something much like **B** on a spectrum analyzer. Note that the harmonics are significantly reduced.

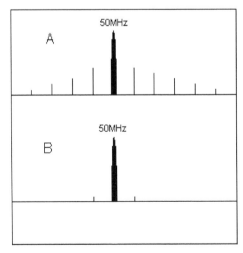

Results of filtering

The actual circuit is not quite as simple as the one shown in the first diagram. Firstly, the output of the oscillator is quite high, so we are going to attenuate it before applying it to the crystal. To do this we, will be using a **pi pad**. It is called this because the arrangement of the resistors looks like the Greek letter pi (π).

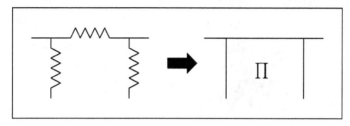

A pi pad

The following diagram is the actual circuit. Resistors **R1** through **R3** make up the pad.

The letters **A** and **B** refer to the signals in the previous figures (*Diagram of a crystal filter* and *Results of filtering*).

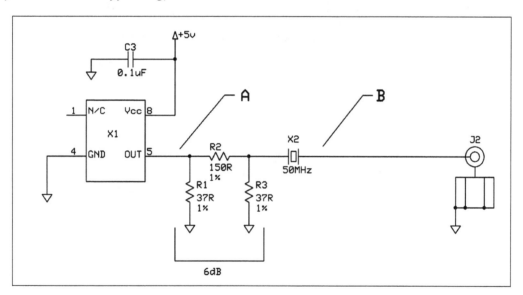

Actual circuit

The pad is also used to trim the output for 0 dBm. I would suggest this level because it is somewhat standard and makes the mathematics easy if you decide to add external attenuators.

The following image is of the actual output from the crystal filter to the 50-ohm load of the spectrum analyzer.

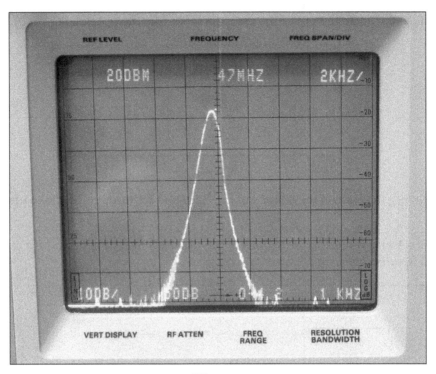

RF output

For those of you who are unfamiliar with spectrum analyzers, I'll explain a few things. The top of the screen (**Ref Level**) is +20 dBm and the scale (**Vert Display**) is 10 dB per division; so two squares down from the top is 0 dBm. This is the output level we are trying to achieve. Each square is divided into five equal parts or 2 dB, so our output signal in this case is about +2 dBm. This measurement was made right at the crystal. Once the board and output connector losses were subtracted, I had an output of exactly 0 dBm.

Why 0 dBm? Because it's a common reference value in RF measurements. Basically, it makes the mathematics easier. If you look at the right-hand side of the screen on my analyzer, you will see a scale marked **LOG dB**. If you set the Reference Level in the **Ref Level** field to 0 dBM and have a 0 dBm output from your circuit, then gain and attenuation measurements are a piece of cake.

I have no way of knowing what the output of your oscillator and crystal combination will be, so you are going to have to calculate your own attenuation pad values.

There are numerous sites on the Internet that will provide you with the formula for doing this, but personally I am lazy so I used the following site to do it for me:

```
http://n9zia.ampr.org/att_pad.main.cgi
```

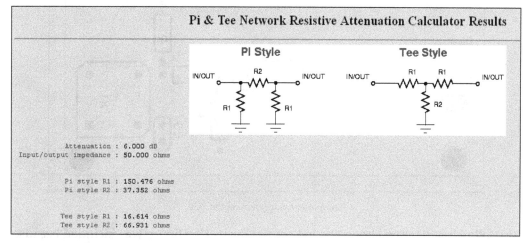

Pi attenuator calculator

The preceding screenshot shows the results for my pad. I simply chose the closest 1-percent value for my resistors. There might be some trial and error involved here, but 6 dB seems to be a good value to start with.

Assembling the PCB

The following diagram shows the component placement on the actual PCB, which is available for download from the Packt site.

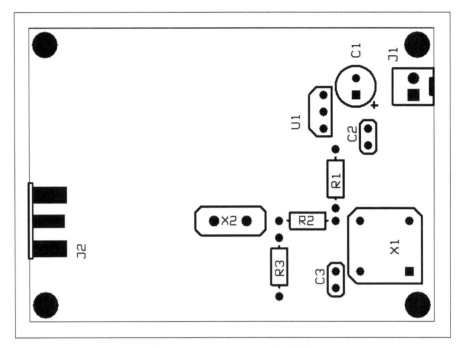

Component placement on the PCB

The following diagram shows the top copper layer. This figure was included because it is important to show the pattern of vias placed on the board. This pattern ensures that there is a good connection between the top and bottom ground planes. If you decide to make your own board, you should do the same.

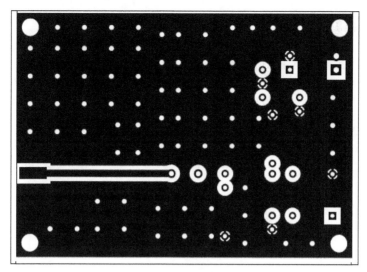

Top copper layer

The following image is of the finished PCB. The original design had two pi attenuators. As it turns out, only one was necessary, so your board will look slightly different. The jumper will not be present.

Finished prototype

You will notice that there is no solder mask or silkscreen on this PCB. This is a common practice of mine when building RF circuitry, because you never know when you will have to add decoupling or a shield.

Going further

What we now have is a calibrated 50 MHz RF source that we will use to test our next two projects. As I mentioned before, the output is 0 dBm. This is just the value I chose. You can set it to any value you like using the attenuator calculator web page. In all projects from now on, I will assume that you are using a 50 MHz 0 dBm source.

You can achieve different output levels by using either a fixed or variable attenuator. The following image shows an example of a fixed attenuator:

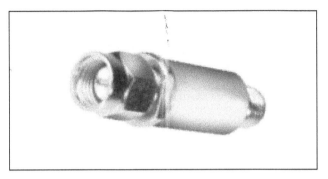

SMA attenuator (copyright Mini-Circuits)

I used an SMA connector on my prototype because of its small size and ease of mounting to the PCB. Theses connectors have a much higher frequency range than we will be using on these boards, so you can use a BNC connector if you like. The following image is of a BNC-type attenuator:

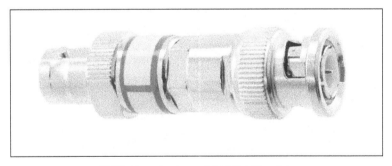

BNC attenuator (copyright Mini-Circuits)

A manual variable attenuator is shown in the following image. If you are lucky enough to either have or borrow one of these, your life will be a lot simpler. Make yourself a couple of coax pigtails and connect it between the output of the oscillator and the crystal. Then just adjust it for the output level you want and read the setting off the side of the attenuator. No trial and error, or at least less, are needed; just enter the setting on the web page and off you go!

Manual attenuators

Programmable attenuator

For this part of the project, we will be using the Mini-Circuits ZX76-15R5-PP+ programmable attenuator. In the following image, you will notice that the attenuator has SMA input and output. That is the other reason I chose an SMA connector for the oscillator. No adapter required!

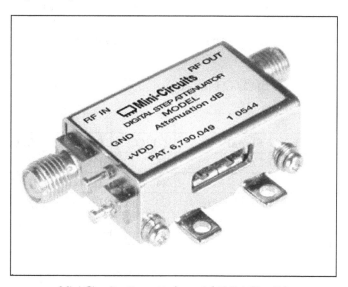

Mini-Circuits attenuator (copyright Mini-Circuits)

The internal construction of the attenuator is actually quite simple. It's a bank of switches that can be opened or closed to remove or add attenuation, respectively. These switches are controlled by 3 V digital logic.

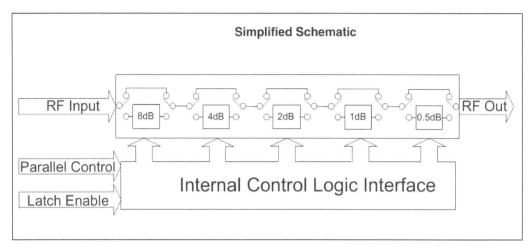

Attenuator internal schematic (copyright Mini-Circuits)

The fact that the device uses 3 V logic makes it easy to interface to our BeagleBone.

BeagleBone I/O pins

The operating system image that comes with this book is supposed to work on either the BeagleBone White board or the newer BeagleBone Black board. The Black board uses some of the I/O for the LCD display and the eMMC chip, thus, in order to make the system hardware work with either version, we must take care not to use any of the I/O used by the Black board.

The following table shows the I/O pins common to both boards.

P9			P8		
GPIO	Pin #	Signal	GPIO	Pin #	Signal
N/A	1	Gnd	45	11	1 dB
	2	Gnd	44	12	2 dB
	3	VDD_3V3	23	13	
	4	VDD_3V3	26	14	4 dB
	43	Gnd	47		8 dB
	44	Gnd	46	16	0.5 dB
	45,46	Gnd	27	17	LE

The attenuator is connected directly to the I/O pins of the BeagleBone, and attenuator power comes from the BeagleBone. The attenuator has only inputs, so power sequencing cannot damage the BeagleBone I/O pins.

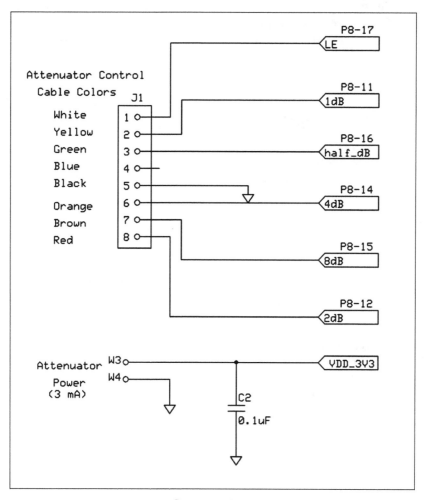

Connector pinout

The signals are brought out to an eight-pin male connector that mates with a female connector installed on the cable purchased separately from Mini-Circuits. You might be interested to know that the same cable is also available with a DB-25 connector. This cable is designed to connect to the printer port on older PCs. Mini-Circuits also provides software to control the attenuator from a Windows program. Depending on your mechanical design, it might be better to order this cable. I'll leave that up to you.

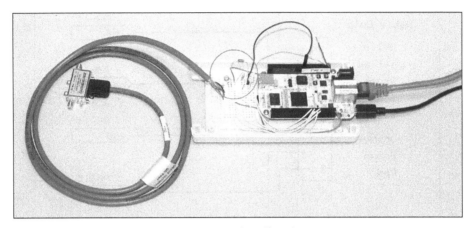

Prototype on breadboard

The preceding image is of the circuit built on a prototype board. The small circuit encircled in red is a test circuit that I built so that I could check the outputs one at a time. I could have just built the interface circuit on one of the many BeagleBone prototype capes available, but I chose to make a custom board that we will be using in the next chapter as well.

The following diagram is of my one-transistor test circuit:

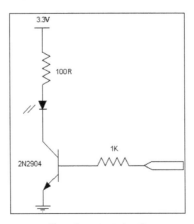

Transistor test circuit

Control software

As with previous projects, I have chosen to use jQuery Mobile to build the GUI. This allows you to control the mobile device if you so choose.

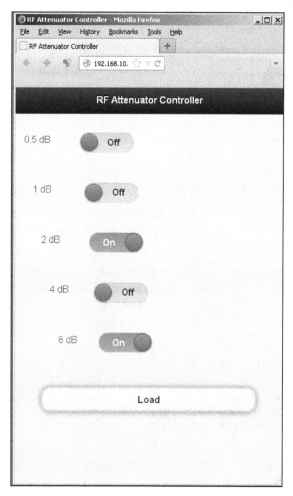

Control interface

The GUI is shown in the preceding screenshot and is very simple to use. Each slide switch corresponds to one control bit of the attenuator. The attenuation is set to *8 dB + 2 dB = 10 dB*.

The code is composed of two parts. The first part is JavaScript that sets up the hardware and then calls the GUI. It also receives and processes commands from the GUI using add-ons called `socket.io` and `bonescript`.

The code is well documented; however, there are a few things that might be of interest to the advanced reader. The following code defines the I/O pins of the BeagleBone and sets up their direction.

```
var outputPin1 = "P8_16"; //0.5dB
var outputPin2 = 'P8_11'; //1dB
var outputPin3 = 'P8_12'; //2dB
var outputPin4 = 'P8_14'; //4db
var outputPin5 = 'P8_15'; //8dB
var outputPin6 = 'P8_17'; //Latch Enable
// configure pins and set all low
b.pinMode(outputPin1, 'out');
b.pinMode(outputPin2, 'out');
b.pinMode(outputPin3, 'out');
b.pinMode(outputPin4, 'out');
b.pinMode(outputPin5, 'out');
b.pinMode(outputPin6, 'out');
```

This is where we use `bonescript` to check the status of each switch on the GUI and react to any change. All the switches are polled in the same manner, except for the **Load** button.

```
io.sockets.on('connection', function (socket) {
  // This is where we check the status of each Switch on the GUI
  socket.on('Output1', function (data) {
  if (data == 'on') {
    //Set the Pin High
    b.digitalWrite(outputPin1,1);
    console.log ("0.5dB On");
  } else if (data == 'off') {
    //Set the Pin Low
    b.digitalWrite(outputPin1,0);
    console.log ("0.5dB Off");
  }
});
});
```

To send this information to the attenuator, press the **Load** button that toggles the **latch enable** (LE) pin of the attenuator and loads the new setting into the device.

Downloading the example code

You can download the example code files from your account at http://www.packtpub.com for all the Packt Publishing books you have purchased. If you purchased this book elsewhere, you can visit http://www.packtpub.com/support and register to have the files e-mailed directly to you.

The **Load** button is processed in a slightly different manner. In this case, the pin is set high and then low. This is done to generate the LE signal required by the attenuator. JavaScript is basically an interoperated language, so it generates a fairly wide pulse as you can see in the following image.

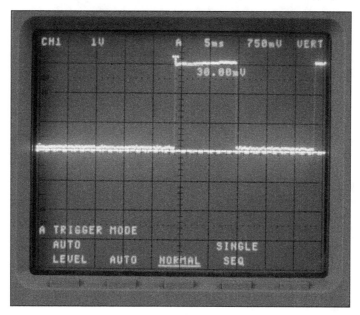

LE signal

The following code toggles the LE pin:

```
socket.on('Output6', function () {
  b.digitalWrite(outputPin6,1);
  console.log("Latch Enable High");
  b.digitalWrite(outputPin6,0);
  console.log("Latch Enable Low");
});
```

The code that generates the actual GUI is written using HTML5 and some style sheet add-ons.

```
<!-- jQuery and jQuery Mobile -->
<link rel="stylesheet"
  href="http://code.jquery.com/mobile/1.3.1/jquery.mobile-
  1.3.1.min.css" />
<script src="http://code.jquery.com/jquery-1.9.1.min.js"></script>
<script src="http://code.jquery.com/mobile/1.3.1/jquery.mobile-
  1.3.1.min.js"></script>
<script src="/socket.io/socket.io.js"></script>
```

Once again, all the switches are handled in the same manner with the exception of the **Load** button.

```
function Switch1(sel){
  if (sel.value == "on") {
    socket.emit('Output1', sel.value);
  } else if (sel.value == "off") {
    socket.emit('Output1', sel.value);
  }
}
```

In the case of the **Load** button, the switch information is first sent as on followed by off to simulate a momentary-action push button.

```
function Switch6(sel){
  socket.emit('Output6', "on");
  socket.emit('Output6', "off");
}
```

The slide switches are defined using the following code:

```
<div data-role="content">
  <div data-role="fieldcontain">
  <label for="Output1">
    0.5 dB
  </label>
  <select name="toggleswitch1" id="Output1" data-theme="b"
  data-role="slider" onchange="Switch1(this);">
  <option value="off">Off</option>
  <option value="on">On</option>
  </select>
  </div>
</div>
```

The code for the **Load** button is considerably simpler:

```
<button type="button" onclick="Switch6()">Load</button>
```

 More information on jQuery and jQuery Mobile can be found at the following links:

http://demos.jquerymobile.com/1.4.2/

http://demos.jquerymobile.com/1.4.2/button/

http://demos.jquerymobile.com/1.4.2/icons/

The following image shows of a 50 MHz, -30 dBm signal that is applied to the attenuator with the attenuation set to 0 dBm.

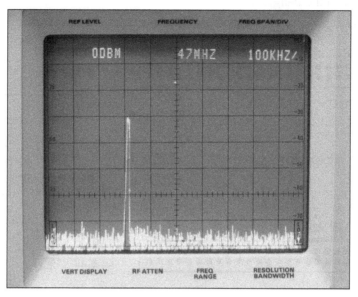

-30dBm signal

The following image is of the same signal with 10 dB of attenuation. Notice that the output is now -40 dBm.

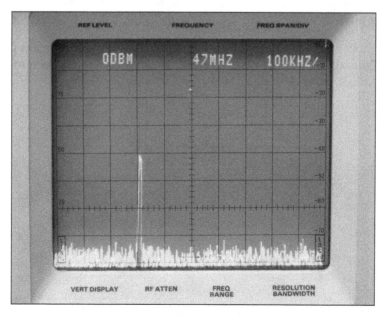

-40dBm signal

The following diagram is of the adapter board that is used to connect the BeagleBone to the attenuator. The same board is also used in the next chapter to connect an RF Power Detector to the BeagleBone.

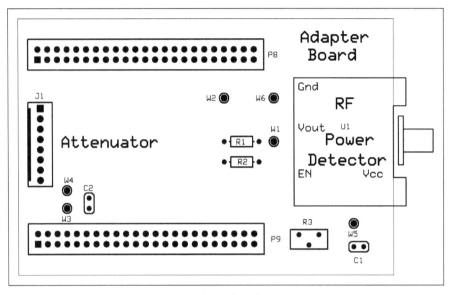

Adapter board

For this chapter, we will only be using the connector **J1** and solder points **W4** and **W3**. The 3.3 V power from the BeagleBone is connected to **W3**, and **W4** is the ground return.

These two connections will have to be attached to the attenuator via a separate cable. The following image shows the cable from Mini-Circuits that has been modified for this project.

Attenuator control cable

Summary

The project in this chapter was probably more challenging than the previous ones, so I hope you were successful and made some new friends in the local ham-radio community.

We learned about crystal filters and how they work. We also learned about crystal oscillators and how they work. The oscillator we used in this project is a simple crystal oscillator. There are two other types of fixed-frequency oscillators that you might have heard of and can substitute if you like.

The first type is called a TCXO, which stands for Temperature Controlled Xtal Oscillator. This oscillator has a built-in heat sinking ability that keeps the crystal at a fairly constant temperature. A crystal's frequency can change slightly with temperature, so this makes a TCXO more accurate than a simple crystal oscillator, and more expensive.

The second type is called an OCXO, which stands for Oven Controlled Xtal Oscillator. As the name implies, the oscillator package contains a heater that keeps the crystal at a constant temperature no matter what the environmental temperature is. These tend to be even more expensive than the TXCO.

In the next chapter, we will use this project to calibrate a digital RF Power meter.

6
RF Power Meter – Hardware

In this chapter, we will be building an RF power meter capable of measuring power levels from approximately +3 dBm to -60 dBm.

For those of you who are not familiar with RF terminology:

- 0 dBm is equivalent to 1 mW or 1/1000 of a watt
- A power level of -30 dBm is equivalent to 1 microwatt or 1/1000,000 of a watt
- A power level of -60 dBm is equivalent to 0.001 microwatts

Basically, the meter is capable of reading extremely low power levels, but requires an external attenuator to measure high power levels. The pros and cons of this feature will become clearer as we go on.

The heart of this project is Linear Technology's LTC5582 RMS RF power detector, as shown in the following image:

Linear Technology power detector

The actual interface to the BeagleBone is quite simple. Power is supplied by the BeagleBone and one of the analog inputs is used to measure the output from the detector. The output voltage swing from the detector is approximately 1.2 volts to 2.4 volts. The 2.4-volt output exceeds the maximum input voltage of 1.8 volts to the DAC; thus, to keep things simple, I divide the output of the detector by two. So we now have an input range of 0.6 to 1.2 volts.

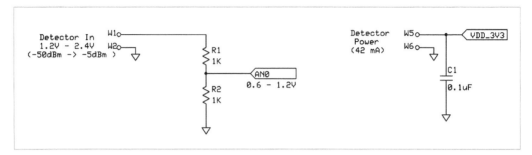

Detector/BeagleBone interface

The graph on page 4 of the Linear Technology data sheet (http://cds.linear.com/docs/en/datasheet/5582f.pdf) shows the detector output voltage versus power at various frequencies. The reader will notice that, below 2700 MHz, the graphs are very similar. In order to measure the higher frequencies, we would have to recalibrate the meter software.

In the previous chapter, we saw the layout of the custom adapter PCB that was used in this project. For the purpose of writing this book, I simply bought a prototype board from Circuitco and used it. You can use it too, if you like.

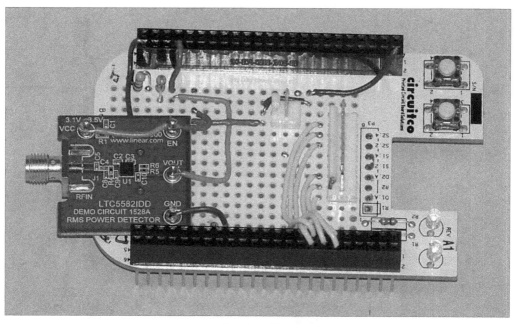

Protoboard top

I simply mounted the demo board on the Circuitco adapter using double-sided tape. The resistor divider is in the upper left of the board. I also installed the 8-pin connector required by the attenuator in *Chapter 5, Calibrated RF Source*, as well as the 3.3-volt power connector. All I did was to remove P3 in order to make room for the 8-conductor cable.

This way, the RF step attenuator can be used with either project. One thing that you should be aware of is that the power detector has a wider bandwidth than the attenuator, so you will have to use discrete attenuators at higher frequencies.

Making power measurements

One of the handy features of this detector is that it gives you an output voltage that is proportional to the RMS input power.

What is RMS?

Find out more about RMS at `http://en.wikipedia.org/wiki/Root_mean_square#RMS_of_common_waveforms`.

What this means to us is that we don't have to calculate the RMS power with our software; the chip does that for us. The schematic of the demo board is shown in the following diagram:

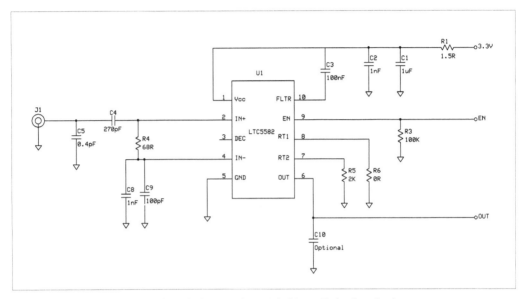

Demo board schematic (copyright Linear Technology Inc.)

The most important features of the schematic are the fact that the input is 50 ohms, single-ended (not double-ended, using a transformer), and the fact that the 3.3 V power is well filtered. This means that we do not have to worry about noise on the 3.3 V power coming from the BeagleBone.

In our case, the EN or Enable pin is simply connected to the 3.3 V rail, because I couldn't see any value in turning the detector on and off. If you want to do this, I suppose you could connect the EN pin to a BeagleBone I/O pin.

Testing and calibration

Now that we have our circuit built, it is time to test and calibrate it. In order to do that, I have provided you with simple JavaScript that you can use to read the input from AN0 on the BeagleBone:

```
var b = require('bonescript');
var inputPin = "P9_39";
loop();
function loop() {
  var value = b.analogRead(inputPin);
  var diff = value-0.667; //0dBm input
  console.log ("Value");
  console.log(value); //DAC Reading
  console.log("Difference");
  console.log(diff); //Difference between
  //current reading and 0dBm
  console.log ("Power");
  console.log((diff/0.007)); // ~ 7% change per dBm
  setTimeout(loop, 1);
}
```

The preceding script outputs various calculations for calibration and debugging purposes. The first variable `value` is the raw input from the DAC. This value is a number between 0 and 1 that represents the percentage of the full-scale reading. So a reading of 0.667 means that the voltage in is 66.7 percent of 1.8 volts, or full scale. This can be confusing, but I managed to determine that a change of 1 dBm in input power results in a 7 percent change in voltage reading by the DAC.

In order to calibrate our software, we will first run the software with a 0dBm power input. We will make note of what the first output is to the console. In my case, it was about 0.667 or 66.7 percent.

Stop the code and enter your number in place of 0.667 in the code. If you increase or decrease the input power by 1 dB, you can confirm that the reading changes by about 7 percent or whatever your number is. You can now modify your code accordingly.

Once we have a reference point, 0 dBm, and a percentage change of 7 percent, we can now calculate the input power by subtracting the 0dBm reading from the current reading and dividing by the change per dBm value. In my case, the change is 7 percent but yours might be slightly different.

If you comment out the debugging information, you should get a real-time reading of the input power to the detector. Out of all the `console.log` statements, the last one does the calculation.

Here we have the script running in the Cloud IDE provided with BeagleBone. The input power is +1 dBm. As you can see in the following screenshot, the software is giving us a reading of **1.06** dBm power:

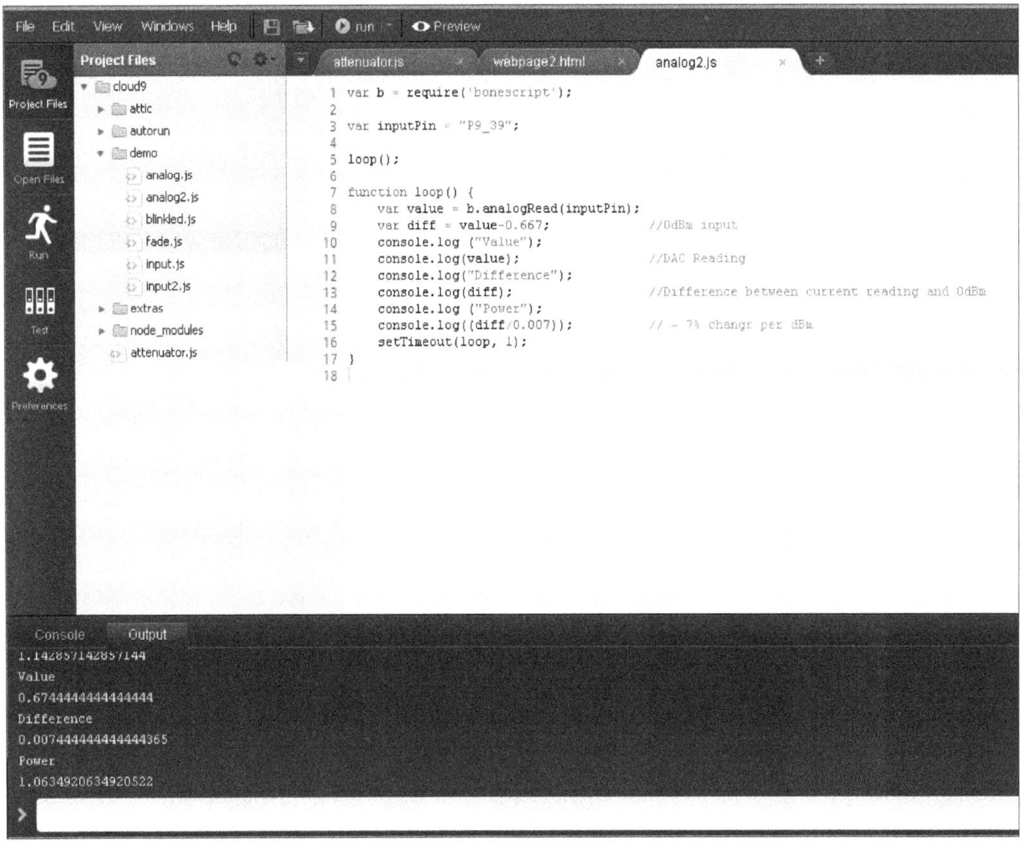

Script running in Cloud9 IDE

You will notice that the readings tend to jump around a bit. Using the custom PCB, and perhaps a filter capacitor from the AN0 input to ground, will help in this situation. I'll leave that refinement to the reader.

Making actual measurements

For those unfamiliar with this type of power meter, there are two types of measurement you can make. The first and by far the easier is the direct-connect method. This is the one I used when doing the calibration. One simply connects the meter to the RF source with a cable. At one point, I had access to high-end equipment and was able to measure the loss in the test cables that I used. For this project, that kind of precision is probably not necessary, except at higher frequencies.

The second type of measurement is off-air measurement. In this case, an antenna and various filters are used. Probably, the single most important component in off-air measurements is the calibrated antenna. These antennas can range in price from tens to thousands of dollars.

The antennas that I will be using in this book are available on eBay for about \$20 to \$30 at `http://stores.ebay.ca/KB5UBE-Engineering/About-KB5UBE-Engineering.html`.

They come with calibration information in the form of antenna factor numbers.

There is a good explanation of antenna factors available from Wikipedia so I won't go into it here.

You can read more at `http://en.wikipedia.org/wiki/Antenna_factor`.

What we really need to do is figure out what the gain of our antenna is at a given frequency. Once again, it is Google to the rescue.

A handy web page for converting antenna factor to antenna gain is at the following link:

`http://rfcalculator.mobi/gain-from-antenna-factor.html`

I am not going to go into the details of RF measurement in this book because there are many good reference books available, and frankly, the math is beyond me and probably beyond many of the readers of this book.

All you Hams and RF engineers out there: if I have offended you, I apologize in advance.

Here, we have a typical setup for off-air measurements. The antenna I am using is a log-periodic antenna. I chose it because it allows me to measure a number of different frequencies with the same antenna. For those of you who are unsure of what a log-periodic antenna is or would like to learn more, I have provided the following link.

 Explanation of what a log-periodic antenna is available here:
http://en.wikipedia.org/wiki/Log-periodic_antenna

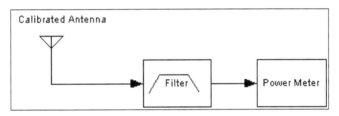

Typical test setup

What you have to do is select a filter for the frequency range that you want to measure. Shown in the following images are typical bandpass filters. In this case, they are connector-connectorized versions from Mini-Circuits. Mini-Circuits also sells solder in module versions of the same filter.

Filter for 2.4 GHz Wi-Fi band

The preceding image shows a bandpass filter for the 2.4 GHz ISM band used by Wi-Fi, Bluetooth, and many cheap video cameras.

5.8 GHz bandpass filter

The filter shown in the preceding image is for the 5.8 GHz band that is used by many cordless phones and other devices. The frequency response of these filters is shown in the following pages.

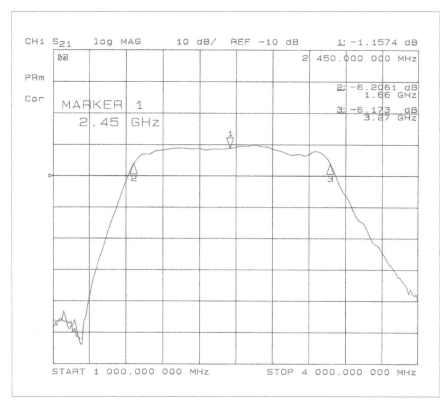

2.4 GHz bandpass filter response

In the preceding graph, you can see that the filter has a fairly sharp roll-off (steep edges), which makes it a reasonable filter for our purposes.

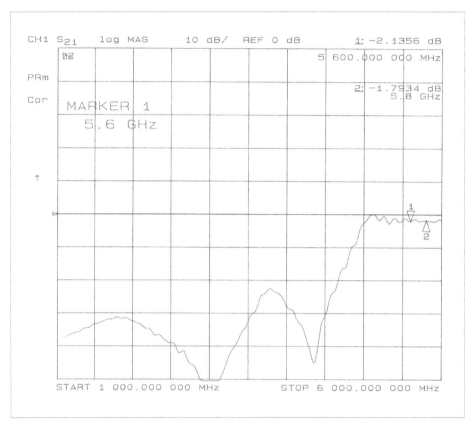

5.6 GHz bandpass filter response

In this case, the network analyzer I had access to only had a measurement range of up to 6 GHz, but you can clearly see the steepness of the filter edge.

The following image shows the antenna I will be using in this book, along with antenna factors at various frequencies. Remember that a log-periodic antenna is a broadband antenna, so we have to specify the antenna factor at various frequencies. In practical applications, I would use the antenna factor closest to the frequency I was measuring.

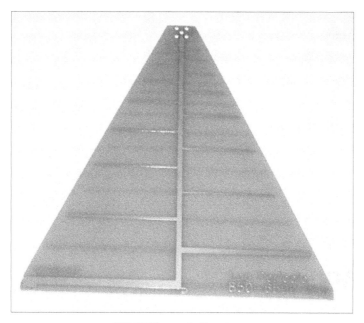

850-6500 log-periodic antenna

Here, we have an antenna that will probably measure any frequency in the ISM bands that you might be interested in, from 900 MHz to 5.8 GHz. An antenna for measuring lower frequencies would be much larger since $L = C / F$, where L is the wavelength, C is the speed of light and F is the frequency in hertz.

A simple rule of thumb I use is that, since the speed of light is 300 meters per microsecond, a 300 MHz signal will have a wavelength of 1 meter. Therefore a quarter-wave, 300 MHz antenna is 250 mm long (one quarter of a meter).

The data sheet for this antenna is available for downloading, but I have included the antenna factor information in the following table as an example of how to calculate antenna gain, given the antenna factor at a given frequency.

Frequency	Antenna Factor	Gain
900MHz	24.0	5.29dBi
1.0GHz	24.2	6.01dBi
1.5GHz	27.5	6.23dBi
2.0GHz	30.1	6.13dBi
2.4GHz	32.0	5.81dBi
3.0GHz	33.0	6.75dBi
6.0GHz	40.0	5.77dBi

The term dBi means dB gain compared to a perfect isotropic antenna. An isotropic antenna is an antenna with a perfectly circular antenna pattern. These antennas do not exist in real life for various reasons. This is simply a term used to define antenna gain.

 Read more on how to make antenna measurements at the following link:
`http://en.wikipedia.org/wiki/DBi#Antenna_measurements`

The takeaway from the measurements in the preceding table is that the gain of the antenna tends to vary a bit over the frequency range of the antenna. For a $30 antenna etched on PCB material, it is pretty good. Certainly, it is good enough for home hobbyists.

Summary

In this chapter, we built and tested the RF Power Meter hardware that was based on the Linear Technology LTC5582 (RMS) power detector. We also wrote some simple software to read one of the analog inputs on the BeagleBone and convert the reading to a dBm power measurement. We briefly discussed making hard-wired and off-air power measurements using various bandpass filters. In the next chapter, we will be developing the BeagleBone software that will be used with this hardware.

RF Power Meter – Software

7

In this chapter, we will be building the development system that I used to write the remaining software. Those of you who would rather just download the image file can skip to the actual software section, but beware that it's an 8-GB file.

Suggested hardware setup

Firstly, let me tell you that I have no financial connection to Special Computing. I just really like their stuff. If you already have a BeagleBone Black board (covered in the previous chapters), then you can purchase only the docking station and cables by contacting them via e-mail or by visiting the following link:

```
https://specialcomp.com/beaglebone/index.htm
```

My development system hardware

Part 1 – installing and configuring the OS

The first step is to get the latest Debian ARM7 image from the following link:

```
http://debian.beagleboard.org/images/bone-debian-7.5-2014-05-14-2gb.
img.xz
```

Simply burn the image to an 8-GB, class 10 microSD card. There are a lot of instructions to do this with various operating systems, so I won't include them here.

Once you have burned the image, install the card in the BeagleBone and power it up. The default username is debian and the password is temppwd.

The easiest way to access the BeagleBone I have found is via SSH, even though the Lapdock has a keyboard. The brief instructions for setting up SSH using PuTTY are explained in the following section.

Setting up PuTTY

After connecting the board to the same network of the host computer, on the development system lapdock open a terminal window and enter ifconfig. The results are shown in the following screenshot:

```
eth0      Link encap:Ethernet   HWaddr c8:a0:30:ac:bf:56
          inet addr:192.168.10.108  Bcast:192.168.10.255  Mask:255.255.255.0
          inet6 addr: fe80::caa0:30ff:feac:bf56/64 Scope:Link
          UP BROADCAST RUNNING MULTICAST  MTU:1500  Metric:1
          RX packets:6918 errors:0 dropped:0 overruns:0 frame:0
          TX packets:1525 errors:0 dropped:0 overruns:0 carrier:0
          collisions:0 txqueuelen:1000
          RX bytes:4104837 (4.1 MB)  TX bytes:142389 (142.3 KB)
          Interrupt:56

lo        Link encap:Local Loopback
          inet addr:127.0.0.1  Mask:255.0.0.0
          inet6 addr: ::1/128 Scope:Host
          UP LOOPBACK RUNNING  MTU:65536  Metric:1
          RX packets:0 errors:0 dropped:0 overruns:0 frame:0
          TX packets:0 errors:0 dropped:0 overruns:0 carrier:0
          collisions:0 txqueuelen:0
          RX bytes:0 (0.0 B)  TX bytes:0 (0.0 B)
```

Result of the ifconfig command

You will notice in the second line of the preceding screenshot that the Internet address is 192.168.10.108. This is the IP address that DHCP assigned to the BeagleBone on my lab's router. Yours will depend on your network settings. I will use this address in my examples from now on.

PuTTY SSH setup screen

The preceding screenshot is of the PuTTY setup screen. All you should have to do is enter the network address of the development system. In my case, it is 192.168.10.108 as I mentioned earlier. Now that we have PuTTY installed and set up, we can continue the installation and setup of our development system:

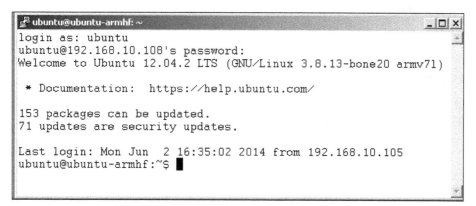

SSH login screen

Setting up root access

For a number of programs to run properly, we will need root access. Normally, this is not a wise privilege to give to a user, so we will do it from a terminal window rather that setting up a root user on the Ubuntu desktop. The following are brief instructions on how to do this, so enter the following commands:

1. apt-get update.

2. sudo passwd root.

3. Enter a root password.

You should now have root access, so you can start another SSH session and log in as a root user.

Expanding the filesystem

There's an excellent tutorial on expanding the filesystem at the following link:

```
http://elinux.org/Beagleboard:Desktops_On_Ubuntu/Debian#Ubuntu_
Precise_On_A_microSD_With_Ubuntu_Desktop
```

However, I will outline the basic steps as follows:

1. First of all we must execute the `fdisk` command on the device containing the Linux root partition. To do this, we enter the following:

   ```
   fdisk /dev/mmcblk0
   ```

2. You will see something similar to the following commands. The sector information will be different on your screen:

   ```
   Command (m for help): p

   Disk /dev/mmcblk0: 8270 MB, 8270118912 bytes
   4 heads, 16 sectors/track, 252384 cylinders, total 16152576
   sectors
   Units = sectors of 1 * 512 = 512 bytes
   Sector size (logical/physical): 512 bytes / 512 bytes
   I/O size (minimum/optimal): 512 bytes / 512 bytes
   Disk identifier: 0x80000000

           Device Boot      Start         End      Blocks   Id
   System
   /dev/mmcblk0p1    *        2048        4095        1024    1  FAT12
   /dev/mmcblk0p2            4096     3751935     1873920   83  Linux

   Command (m for help):
   ```

3. Next, enter d to delete a partition and then enter 2 for partition 2:

   /dev/mmcblk0p2

4. Now create a new partition by entering n, p, and then 2.

5. You should hit *Enter* to have your default start sector used.

6. Hit *Enter* to use the entire card.

   Now it's time to write the information to the microSD card:

7. In order to do that, we enter w to write out the changes as shown in the following command:

   ```
   Command (m for help): w


   The partition table has been altered!


   Calling ioctl() to re-read partition table.


   WARNING: Re-reading the partition table failed with error 16:
   Device or resource busy.
   The kernel still uses the old table. The new table will be used at
   the next reboot or after you run partprobe(8) or kpartx(8)
   Syncing disks.
   ```

8. Reboot the system using the following command:

   ```
   # reboot
   ```

9. Now it's time to actually resize the filesystem by entering the following:

   resize2fs /dev/mmcblk0p2

10. If we enter the df command, we will see the following:

    ```
    Filesystem        1K-blocks     Used Available Use% Mounted on
    /dev/mmcblk0p2    7674968 501400   6845508    7% /
    devtmpfs           253772       4    253768    1% /dev
    none                50784     276     50508    1% /run
    none                 5120       0      5120    0% /run/lock
    none               253916       0    253916    0% /run/shm
    /dev/mmcblk0p1       1004     472       532   48% /boot/uboot
    ```

Now that we have resized the filesystem, it's time for the next step: installing the Ubuntu desktop.

Installing the Ubuntu desktop

The following are the steps to install the Ubuntu desktop on our system:

1. First we must update the file database with the following commands:

    ```
    sudo apt-get update
    ```

2. Then we will update to the latest copies of everything in the root file system with the following commands:

    ```
    sudo apt-get upgrade
    ```

3. Once the software has finished downloading and installing, it's time to install the desktop:

    ```
    sudo apt-get install Ubuntu-desktop
    ```

Part 2 – Installing the additional software and dependencies

Follow these steps to install the additional dependencies:

1. Install TightVNC Server by entering the following commands:

    ```
    apt-get install tightvncserver
    ```

    Enter vncserver to start the server. The first time you run VNC server, you will see the following:

    ```
    root@ubuntu-armhf:~# vncserver
    You will require a password to access your desktops.
    Password:
    Verify:

    Would you like to enter a view-only password (y/n)? n
    xauth:  file /root/.Xauthority does not exist
    New 'X' desktop is ubuntu-armhf:1
    Creating default startup script /root/.vnc/xstartup
    Starting applications specified in /root/.vnc/xstartup
    Log file is /root/.vnc/ubuntu-armhf:1.log
    ```

    In future, you can just type vncserver in the terminal or add it to a startup script.

2. Install the Emacs editor. This is optional:

```
apt-get install emacs
```

3. Install Build Essentials:

```
apt-get install -y build-essential g++ curl libssl-dev apache2-
utils git libxml2-dev
```

Installing the Device Tree Compiler

Build the Device Tree Compiler from source code using Robert Nelson's script:

```
wget -c https://raw.github.com/RobertCNelson/tools/master/pkgs/dtc.sh
chmod +x dtc.sh
./dtc.sh
```

Installing Derek Molloy's Device Tree Source (Optional)

To install the Device Tree Source, do the following steps:

1. Install the Device Tree source code from Derek Molloy using the following command:

```
git clone https://github.com/derekmolloy/boneDeviceTree.git

cd ~/boneDeviceTree/overlay
```

2. Edit the device tree source file:

```
 nano DM-GPIO-Test.dts
```

3. Compile the source code using the Make script provided:

```
./build
```

4. Copy the object file to the firmware directory:

```
cp DM-GPIO-Test-00A0.dtbo /lib/firmware/
```

5. To check that it is present:

```
Cd /lib/firmware
Ls
```

6. Apply the overlay to the Kernel:

```
Echo DM-GPIO-Test > $SLOTS
```

Installing Node.js

To install Node.js, run the following commands:

```
wget http://nodejs.org/dist/v0.8.25/node-v0.8.25.tar.gz


tar -zxvf node-v0.8.25.tar.gz


cd node-v0.8.25


./configure


make


make install
```

Setting up external storage

To set up external storage, add an entry to /etc/fstab that automatically mounts the drive at boot time, as shown in the following commands:

```
proc /proc proc defaults 0 0
/dev/mmcblk0p2          /               auto    errors=remount-ro   0   1
/dev/mmcblk0p1          /boot/uboot     auto    defaults            0   0
/dev/sda1               /media/disk1    vfat    auto,umask=0 0 0
```

Then reboot. This will mount the USB memory stick each time the BeagleBone is booted.

If you type mount after the reboot, you will see something similar to the following commands:

```
# mount


/dev/mmcblk0p1 on /boot/uboot type vfat (rw)
/dev/sda1 on /media/disk1 type vfat (rw,umask=0)
```

External filesystem

Now that we have the external storage set up, we can continue with the software installation.

Installing BoneScript

To install BoneScript, do the following steps:

1. Enter cd /home/debian.

2. Enter mkdir local to create a local directory.

3. Enter cd /home/debian/local to switch to your local directory.

4. Enter npm install bonescript to install BoneScript.

Installing Cloud9

Without leaving the `/home/debian/local` directory, enter the following commands:

```
git clone https://github.com/ajaxorg/cloud9/
cd cloud9
npm install
chmod 777 .sessions
cd ~/cloud9/node_modules
```

Installing socket.io

The command to install `socket.io` is as follows:

```
npm install socket.io
```

Modifying the profiles in root and Debian

For user (Debian): `nano /etc/profile`

```
PATH=$PATH:/home/debian/local/cloud9/:/home/debian/local/cloud9/bin/


export PATH


export NODE_PATH=/usr/local/bin/cloud9/node_modules


# Required For Device Tree Overlays
export SLOTS=/sys/devices/bone_capemgr.9/slots
export PINS=/sys/kernel/debug/pinctrl/44e10800.pinmux/pins
```

For root: `nano /root/.bash_profile`

```
PATH=$PATH:/home/debian/local/cloud9/:/home/debian/local/cloud9/bin/


export PATH


export NODE_PATH=/usr/local/bin/cloud9/node_modules
# Required For Device Tree Overlays
export SLOTS=/sys/devices/bone_capemgr.9/slots
export PINS=/sys/kernel/debug/pinctrl/44e10800.pinmux/pins
```

Our new development system file structure should look like the following figure:

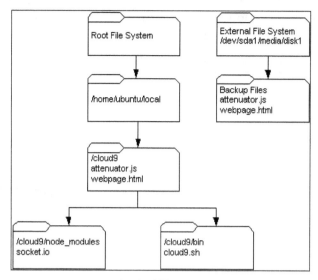

Finished filesystem

To run the Cloud9 **Integrated Development Environment** (IDE), log in to the root via SSH and enter `cloud9.sh -l 0.0.0.0`. You should get a message in your terminal window similar to the following screenshot:

```
root@ubuntu-armhf:~# cloud9.sh -l 0.0.0.0
make: Nothing to be done for `worker'.
Linux ARM
connect plugin start
Connect server listening at http://0.0.0.0:3131
IDE SERVER PLUGIN:    auth
IDE SERVER PLUGIN:    git
IDE SERVER PLUGIN:    gittools
IDE SERVER PLUGIN:    hg
IDE SERVER PLUGIN:    npm
IDE SERVER PLUGIN:    filelist
IDE SERVER PLUGIN:    search
IDE SERVER PLUGIN:    revisions
IDE SERVER PLUGIN:    settings
IDE SERVER PLUGIN:    shell
IDE SERVER PLUGIN:    state
IDE SERVER PLUGIN:    watcher
IDE SERVER PLUGIN:    node-runtime
IDE SERVER PLUGIN:    npm-runtime
IDE SERVER PLUGIN:    python-runtime
IDE SERVER PLUGIN:    apache-runtime
IDE SERVER PLUGIN:    ruby-runtime
IDE SERVER PLUGIN:    php-runtime
Started '/home/ubuntu/local/cloud9/configs/default'!
IDE server initialized. Listening on 0.0.0.0:3131
```

Cloud9 running on BeagleBone

Using the RF power meter software

If we open another SSH session and look at the $SLOTS environment before running the attenuator software, we will see the following:

```
root@ubuntu-armhf: ~                                                    _|□|×|
root@ubuntu-armhf:~# cat $SLOTS
 0:  54:PF---
 1:  55:PF---
 2:  56:PF---
 3:  57:PF---
 4:  ff:P-O-L Bone-LT-eMMC-2G,00A0,Texas Instrument,BB-BONE-EMMC-2G
 5:  ff:P-O-L Bone-Black-HDMI,00A0,Texas Instrument,BB-BONELT-HDMI
 6:  ff:P-O-L Override Board Name,00A0,Override Manuf,DM-GPIO-Test
root@ubuntu-armhf:~#
```

Before running the attenuator software

If we then run the software in Cloud9 IDE and again check the $SLOTS environment, we will see that a number of additional slots have been added by the software:

```
root@ubuntu-armhf:~# cat $SLOTS
 0:  54:PF---
 1:  55:PF---
 2:  56:PF---
 3:  57:PF---
 4:  ff:P-O-L Bone-LT-eMMC-2G,00A0,Texas Instrument,BB-BONE-EMMC-2G
 5:  ff:P-O-L Bone-Black-HDMI,00A0,Texas Instrument,BB-BONELT-HDMI
 6:  ff:P-O-L Override Board Name,00A0,Override Manuf,bspm_P8_16_f
 7:  ff:P-O-L Override Board Name,00A0,Override Manuf,bspm_P8_11_f
 8:  ff:P-O-L Override Board Name,00A0,Override Manuf,bspm_P8_12_f
 9:  ff:P-O-L Override Board Name,00A0,Override Manuf,bspm_P8_14_f
10:  ff:P-O-L Override Board Name,00A0,Override Manuf,bspm_P8_15_f
11:  ff:P-O-L Override Board Name,00A0,Override Manuf,bspm_P8_17_f
12:  ff:P-O-L Override Board Name,00A0,Override Manuf,cape-bone-iio
root@ubuntu-armhf:~# ▊
```

Attenuator.js software running

The additional slots that configure the IO pins on connector P8 are added at the bottom. The JavaScript also added an override for the analog input AN0.

The next two screenshots show the code running in Cloud9 IDE. I have separated the code window from the HTML window so as to make the text more readable. When you run the IDE, the two windows will appear as one. The following screenshot shows the code window:

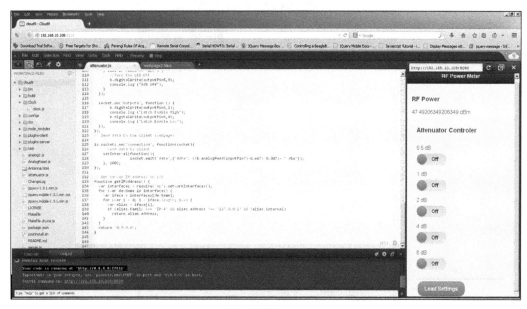

Cloud9 IDE screen

Beneath the code window is the output window that displays messages that would normally appear on your terminal device. Some are messages from the IDE and others are the messages you send via the `console.log` statement.

Output window

The web page (HTML) display is shown in the following screenshot. The icons in the upper right corner allow you to refresh, open a separate browser page, and close the window, respectively.

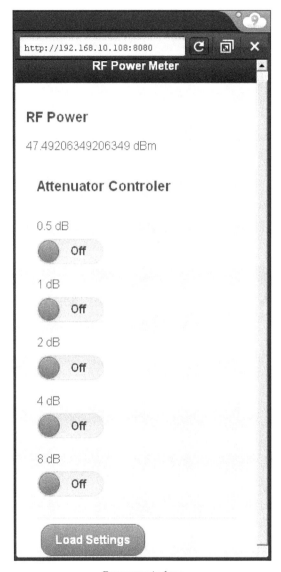

Browser window

 Note that the IP Address shown in the browser box is **192.168.10.108:8080**, the BeagleBone Black board address plus the 8080 port. This is the port referenced in the JavaScript.

The RF detector is not connected, so the RF power measurement is meaningless. It's just for show.

Part A – JavaScript

There are actually two parts of the RF Power Meter software: the JavaScript portion and an HTML web document (called by the JavaScript portion). What follows is a brief description of JavaScript and what its various sections do. This is not intended to replace the inline documentation that is part of the code.

The first thing we do is to define some variables and set up the port on which the application will be listening. In our case, the port number is 8080:

```
var app = require('http').createServer(handler);
var io = require('socket.io').listen(app);
var fs = require('fs');
var b = require('bonescript');

app.listen(8080);

console.log('Server running on: http://' + getIPAddress() + ':8080');
```

Next, we assign variables to the physical pins on the BeagleBone connector. You will notice that inputPin7 is assigned a physical pin number but not configured in the following code, unlike the other pins. This is because the analog input is accessed in a slightly different fashion, as you will see shortly:

```
var outputPin1 = "P8_16";        //0.5dB
var outputPin2 = 'P8_11';        //1dB
var outputPin3 = 'P8_12';        //2dB
var outputPin4 = 'P8_14';        //4db
var outputPin5 = 'P8_15';        //8dB
var outputPin6 = 'P8_17';        //Latch Enable
var inputPin7 = 'P9_39';         //Analog input AN0

// configure pins and set all low
b.pinMode(outputPin1, 'out');
b.pinMode(outputPin2, 'out');
b.pinMode(outputPin3, 'out');
b.pinMode(outputPin4, 'out');
b.pinMode(outputPin5, 'out');
b.pinMode(outputPin6, 'out');
```

Now that we have the pins and variable configured, it's time to load the web page, which will be our user interface. One of the reasons that I chose JavaScript and HTML is that I can access the instrument from a smartphone or tablet. Basically, the code loads the web page and prints an error message if the page is not accessible:

```
function handler (req, res) {
  if (req.url == "/favicon.ico"){ // handle requests for favico.ico
  res.writeHead(200, {'Content-Type': 'image/x-icon'} );
  res.end();
  console.log('favicon requested');
  return;
  }
  fs.readFile('webpage2.html', // load html file
  function (err, data) {
    if (err) {
      res.writeHead(500);
      return res.end('Error loading webpage.html');
    }
    res.writeHead(200);
    res.end(data);
  });
}
```

From now on, what we have to do is check the status of every switch icon on the GUI and perform the requested function on the BeagleBone. Basically, we will set an I/O pin either high or low, based on the message received by socket.io. The code is the same for all six switches, so it is not shown here:

```
io.sockets.on('connection', function (socket) {

socket.on('Output1', function (data) {
    if (data == 'on') {
        //Set the Pin High
        b.digitalWrite(outputPin1,1);
        console.log ("0.5dB On");
    } else if (data == 'off') {
        //Set the Pin Low
        b.digitalWrite(outputPin1,0);
        console.log ("0.5dB Off");
    }
  });
```

Analog data is handled differently, because it travels through `socket.io` in the opposite direction. Rather than the web page (client) sending data to the BeagleBone (server), the BeagleBone reads the analog input, formats the data, and sends it to the web page. It does so every 100 milliseconds:

```
io.sockets.on('connection', function(socket){
    //send data to client
    setInterval(function(){
                socket.emit('date',{'date': ((b.analogRead(inputP
in7)-0.667)/0.007)+ " dBm"});
    }, 100);
});
```

What the preceding highlighted code does is to first read AN0 using the `analogRead()` function. It then subtracts a constant that I measured earlier. This is the reading that I measured with 0 dBm at 50 MHz. Your number might be different, but it should be close. You will know you are close when, with 0 dBm, the display reads about 0 dBm. There is a certain amount of noise in the system, so you will never get an exact 0 dBm reading with 0 dBm input. This is, however, a good test of how well your hardware has been constructed.

Once we have the difference between the current input power and 0 dBm measured, the next step is to calculate the actual input power. From the datasheet provided by Linear Technology, I calculated that there is approximately a change of 7 percent in output voltage for every 1 dBm of input power. This calculation was necessary because the `analogRead()` function produces a number between 0 and 10, with 0 meaning 0 volts and 10 meaning 1.8 volts.

The output of the detector swings between 0 and 2.4 volts. For that reason, a 2:1 voltage divider was used to protect the input of the BeagleBone. Unfortunately, this means that we lose about one third of our dynamic range. There's probably some way of scaling the input using an amplifier, but resistors don't add any noise to the system.

Part B – HTML code

The other part of the software is the HTML page, which is called by JavaScript. The following is a description of the web page and how the various scripts work.

The first thing we have to do is open a socket to the server so that we can send data back and forth:

```
<!-- Code for socket.io  -->

<script>
var socket = io.connect();
// Send data through socket

function Switch1(sel){
        if (sel.value == "on") {
            socket.emit('Output1', sel.value);
        } else if (sel.value == "off") {
            socket.emit('Output1', sel.value);
        }
    }
```

The preceding code opens a socket to the server. The `Switch1` function checks the state of the first toggle switch and sends a message via `socket.io` to the server (BeagleBone), where it is interpreted by the JavaScript as we saw earlier. The code is the same for five of the six switches. The first five switches are *virtual* toggle switches and therefore have an `on` and an `off` position. The sixth switch is a pushbutton, so the code for it is slightly different:

```
function Switch6(sel){

            socket.emit('Output6', "on");
            socket.emit('Output6', "off");
    }
```

There is no decision to be made; the code sends an `on` command followed by an `off` command. This generates a positive-going pulse on the pin of the BeagleBone.

The only remaining part of the script is the code that receives data from the server via `socket.io` and makes it available on the web page:

```
socket.on('date', function(data){
        $('#date').text(data.date);
    });
```

The rest of the code is used to draw various elements on the webpage. The first few lines implement the headers. The highlighted code is what displays the processed analog data on the web page:

```
<body>
<!-- Home -->
<div data-role="page" id="page1">
    <div data-theme="a" data-role="header">
        <h3>
            RF Power Meter
        </h3>
    </div>

    <div data-role="content">
        <h3>
            RF Power
        </h3>
        <div id = "date" /div>
    </div>

        <div data-role="content">
            <h3>
                Attenuator Controller
            </h3>
```

The next few modules draw the various switches and label them. When the object is left-clicked, the value changes between on and off, and the information is sent to the server:

```
<div data-role="fieldcontain">
            <label for="Output1">
            0.5 dB
            </label>
            <select name="toggleswitch1" id="Output1" data-theme="b"
data-role="slider" onchange="Switch1(this);">
                <option value="off">Off</option>
                <option value="on">On</option>
            </select>
```

Again, the exception to this is the pushbutton:

```
<button type="button" onclick="Switch6()">Load Settings</button>
```

There is nothing to change in the GUI, so it simply sends a `switch6()` event to the server.

Summary

In the first part of this chapter, we built our development system and configured the Cloud9 IDE so that it could be run locally on our BeagleBone.

In the second part of the chapter, we enhanced our GUI so that, in addition to controlling the programmable attenuator, it also measured and displayed the RF power.

Once again, the GUI uses the mobile JavaScript add-on, so that the attenuator and power meter can be controlled from a tablet or smartphone.

8
Creating a ZigBee Network of Sensors

In this chapter, we will be creating a star-shaped network of sensors based on the 802.15.4 or ZigBee protocol. We will see how to configure the ZigBee module using Windows-based tools. We will interface the ZigBee module with alarm system boards and test the setup.

It is important to note that XBee modules come in two different types. Not only do they have different features, but they also cannot communicate with one another. In order to give the Series 2 devices mesh network capability, some of the features of the Series 1 modules had to be left out. Most notably, I/O line passing is only available in the Series 1 modules. However, the "pro" versions can communicate with devices in their own series. They just have different antenna connectors and more output power.

 You can find more information related to mesh networking at http://en.wikipedia.org/wiki/Mesh_networking.

Part 1 – setting up the ZigBee modules

Follow the steps to easily set up the ZigBee modules:

1. The first thing we have to do is to install the XCTU software from the Digi website at http://www.digi.com/support/productdetail?pid=3352&type=utilities.

2. The next thing we have to do is to install the first ZigBee module in what is called a "discovery board." This is an adapter that allows us to talk to the module using a virtual serial port.

3. You will need the FTDI USB drivers installed before connecting the discovery board to your computer. These drivers are available at `http://www.ftdichip.com/FTDrivers.htm`.

When you first plug in the discovery board, you will get a **New hardware found** message and your operating system will install a virtual serial port. Depending on your operating system, it will show up in your device manager as "USB Serial Port" and your OS will assign it a com number. In my case, it was assigned com 9. The following image is of my development system hardware:

ZigBee discovery board

Here we have a SparkFun XBee Series 1 board (`https://www.digikey.com`, Catalog#: XBP24-ASI-001-ND) installed on an XBee explorer board (`http://www.robotshop.com/ca/en/sfe-xbee-explorer-usb.html`). I have also installed two LEDs and two push-button switches. The right-most LED flashes indicate whether data is being sent or received. The LED beside it indicates whether or not the module is sleeping. This LED serves as a power indicator, because the module never goes to sleep. The two pushbuttons are used to reset the module after changing the firmware, and the commission (**COMM**) button is used to place the module in various special modes. The functions of these buttons will be described later on. To start the XCTU development package, double-click the icon that was installed by software installer:

1. When we start the XCTU software, the first thing we will be asked to do is to select a serial port for the software to scan for connected ZigBee modules.

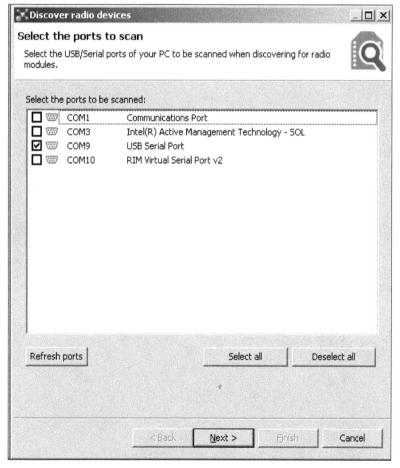

Select a port to scan

2. The next thing we have to do is configure our virtual com port so that it can communicate with the discovery board.

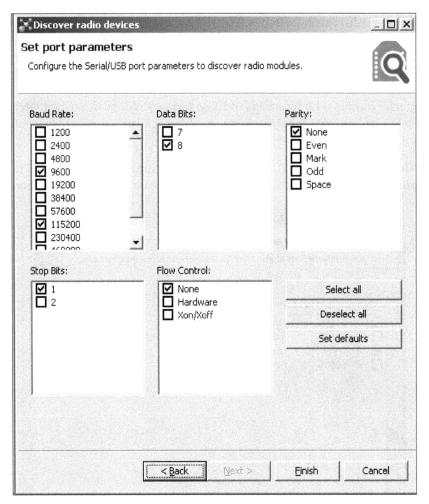

Configure the port

3. The final step is to tell the Windows software to search for ZigBee modules.

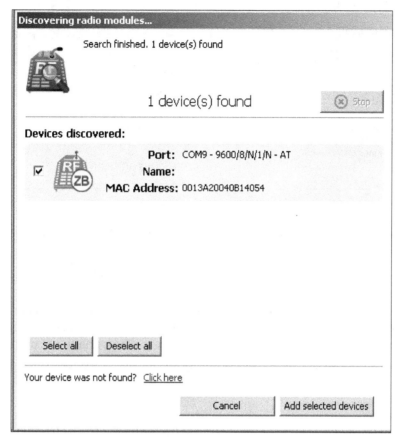

Scan for modules

Once you select the device, the workbench software will query the ZigBee module for its current settings.

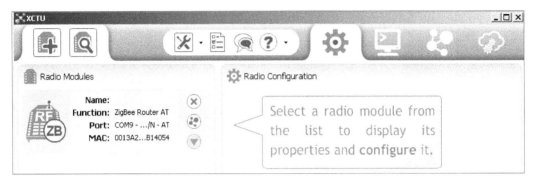

Select a module to program

So, now that we have the hardware and software installed, it is time to start configuring our ZigBee modules:

1. As the preceding screenshot says, just select our module and it is time to play around a bit. The following screenshot shows settings for the XBee pro module, which we will be installing in our alarm panel. This is because it has a higher power output than the other modules and a reverse-polarity SMA connector that allows us to connect an external antenna to the module.

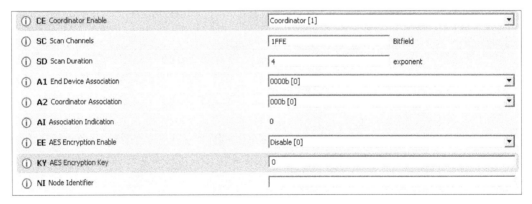

Enabling the Coordinator function

2. Next, we have to set up the I/O settings for this module. The Coordinator module has three I/O pins programmed as outputs, two of which will be used as digital I/O passing outputs.

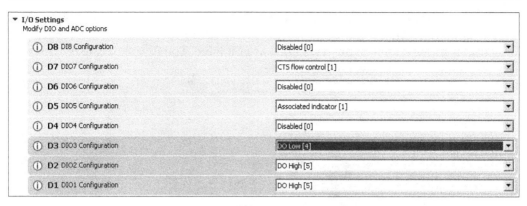

I/O settings

3. In order to enable digital I/O line passing on the Coordinator module, we have to change the I/O input address to **FFFF**.

Input address setup

4. Now that we have everything set up, it is time to write the new settings to the module.

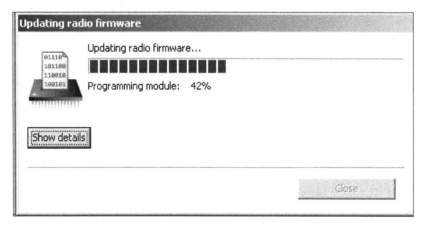

Module programming

5. Now that we have the Coordinator module set up, it is time to program some end units. All the end units are programmed similarly, so you can repeat this process for as many end units as you have. The first step is to set up the destination address for the end unit.

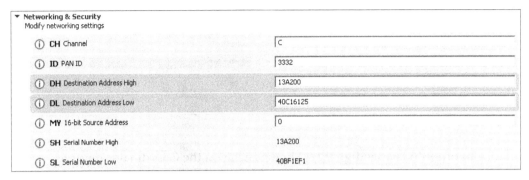

Destination address setup

6. The destination address is the 64-bit address of the Coordinator module. This is where the end units will send their I/O packets and digital I/O line passing information. You will recall that we enabled I/O pins **DI01** and **DI02** as outputs on the Coordinator module. What we do to enable digital I/O line passing is enable the same I/O pins on the end units as inputs.

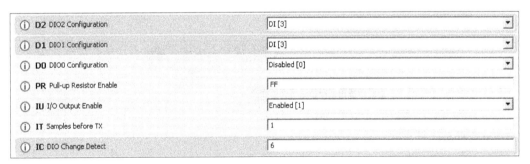

I/O pin configuration

There are two ways in which the Coordinator module can detect a change in the inputs on one of the end (remote) units. It can either continuously ask each unit to sample its inputs, or the units can generate a local interrupt when an I/O line changes state. This second method is much more efficient, and that is the method we will be using. In order to do that, we have to set up what is called an I/O mask. This tells the module which I/O pins to monitor for any change.

There are 8 input output pins, so the mask is 8 bits wide. In our case, we are using I/O pins DIO1 and DIO2, so the mask is: `0000 0110`, or `0x06` in hexadecimal form.

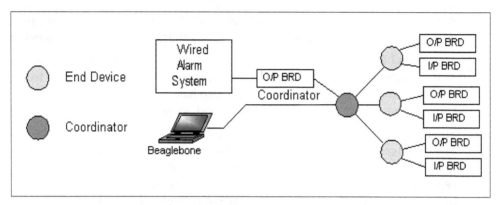

XBee alarm system

The preceding diagram shows the network we'll be building in the remaining portion of this chapter. An alarm input board and an output board are connected directly to the end units. An alarm system output board can also be connected to the Coordinator module so that it can signal the wired alarm system if an event occurs in the wireless portion of the alarm system.

The network topology we're using is called a star topology. This is because, unlike a mesh network that has a mixture of coordinators, routers, and end units, the star configuration has only a coordinator and end units:

- The star topology was chosen for this book because of its simplicity from a software perspective. The reader should be aware that the XBee modules are pin-compatible between series. What this means is that, should the reader wish to implement a series 2 network at a later date, all that it would require is exchanging the series 1 modules for series 2 modules (and a software change, of course).

- The other reason that a star topology was chosen is that it is only in this configuration that I/O line passing is a possibility.

- I/O line passing does not require us to decode any of the receive packets as we would have to with the mesh network topology. The downside to this is that we don't have the robustness (fault tolerance) of a mesh network because XBees cannot "route" (pass on) packets not intended for them.

- In a mesh network, we have three types of XBee devices. We have at least one Coordinator, routers, and end units. A coordinator does the same job in both topologies. It is usually connected to the host computer (in our the BeagleBone) and passes messages to the rest of the network. In series 2 mesh networks, it can also remotely configure the other devices on the network.

- The second type of device is a router. This type acts almost the same as the router connected to your computer at home. Its job is to relay packets to and from the other XBees in the network. It also has I/O of its own thus if the packet is intended for it, then it can respond accordingly.

- The third type is called an end unit. This unit is found at the end of a branch. The only reason to program an XBee as an end unit is so that it can be put to sleep and conserve power. It is actually a kind of napping because it can be programmed to wake up at regular intervals and check for messages (packets), or it can wake when it receives a packet intended for it.

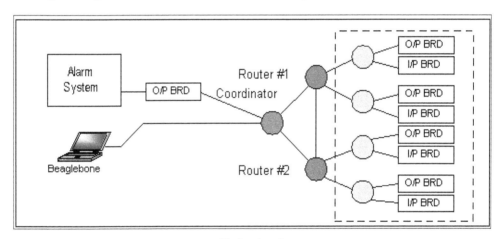

Mesh network

Part 2A – the network hardware

Now that we have the XCTU workbench installed and tested, we can discuss the network we'll be building. There will be two different types of ZigBee devices on the network, from a software perspective. From a hardware perspective, all the units are the same.

The following table shows the standard XBee pin numbers and their functions:

Pin Number	Signal Name	AT Command	Functions
1	Vcc	N/A	3.3V power
2	Tx – Out	N/A	Transmit
3	Rx – In	N/A	Receive
4			
5			
6			RSSI – PWM
7			
8			
9		None	Sleep control
10		N/A	Power supply ground
11		D4	
12		D7	CTS or DIO7
13	On	None	On/Sleep indicator
14	VREF	None	Reference voltage for A/D converters
15	DIO5	D5	Association indicator
16	DIO6	D6	RTS
17	DIO3/AD3	D3	Analog input 3
18	DIO2/AD2	D2	Analog input 2
19	DIO1/AD1	D1	Analog input 1
20	DIO0/AD0	D0	Commissioning button

A ZigBee device can be configured as one of two types. A coordinator is a device in charge of setting up and managing the network and the traffic flow. The other type is called an end unit. As the name suggests, it is at the end of one of the branches of the network tree. It can only receive and process information sent specifically to it.

XBee modules

In the preceding image, one of the end unit (remote) modules is shown on the left and the coordinator is on the right.

There are two types of XBee adapter boards. Here we have the adapter board that will provide the interface between the XBee module and the alarm panel boards, in the end (remote) unit:

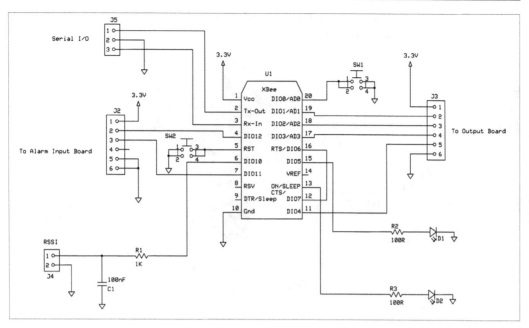

XBee alarm system adapter board

In addition to providing the interface signals, the adapter board will also provide 3.3-volt power to the alarm boards. This board will be mounted on top of the alarm boards in end (remote) units. The 3.3-volt power is supplied by a low-dropout 3.3-volt regulator that is capable of supplying 1-ampere current to the various loads.

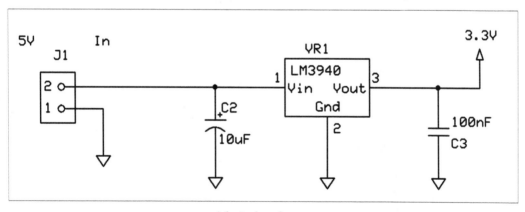

Adapter board power

In the following image, we see the three boards connected together and ready for final assembly in the remote unit:

Remote XBee alarm module assembly

The alarm system consists of two types of boards. The first board is the interface to the alarm system sensors. This board has two outputs. They are connected to DIO11 and DIO12. The second board is a general-purpose output board that provides four optically isolated outputs for controlling external devices. The input to this board is connected to pins DIO1 through DIO4.

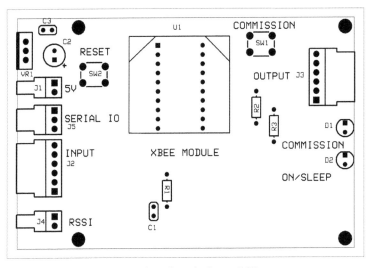

XBee: alarm board adapter PCB

The adapter PCB shown in the preceding diagram has a number of important features. The **COMISSION** button is included for hardware compatibility with series 2 modules, and is not used by series 1 modules.

The **RESET** button is required when changing the firmware from API mode to AT mode. There are also two LED indicators. The On/Sleep indicator should be on when you are trying to communicate with the module. If it is off, pressing the reset button should wake up the module. The commission LED will flash at various rates, depending on what is happening at the time. If the module is programmed and associated with a network, the LED should flash on and off at about a 1 Hz rate.

The other XBee adapter board is the one that mounts on the BeagleBone. The BeagleBone can act as a standalone wireless alarm system, with the capability of signaling a wired alarm system of your choice, or the wired alarm system described in my previous book. The alarm output of the first board is an opto-isolator that can drive a relay. This relay can be connected to either a siren in the stand alone case, or to the zone contacts of a commercial alarm system.

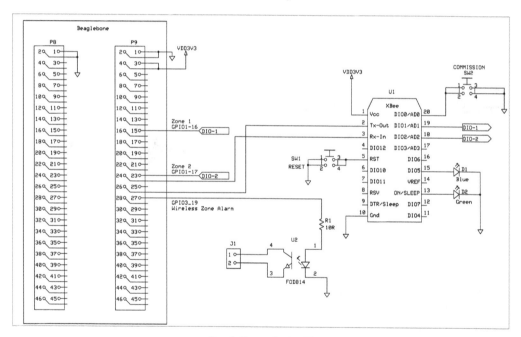

BeagleBone adapter #1

Most of the connections are the same with the exception of the opto-isolator (**U2**) as mentioned earlier and the lack of a 3.3 V regulator. In the latter case, the module receives power from the BeagleBone, so the regulator is not required. The reader will notice that the adapter PCB extends past the edge of the BeagleBone PCB. This was done so as to get the XBee module as far from the BeagleBone CPU as possible in order to limit the amount of RF interference between the module and the BeagleBone.

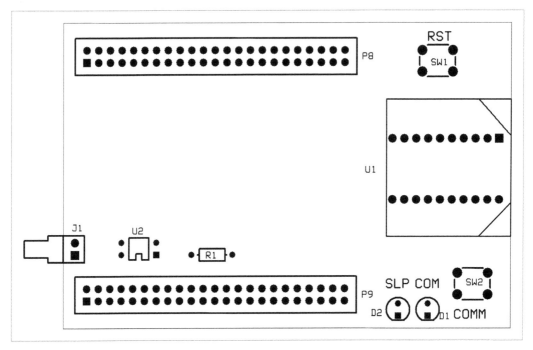

BeagleBone remote adapter PCB

The ground plane has also been removed from under the XBee module for the same reason.

The other PCB that I have provided with this book allows the buyers of my first book (http://www.packtpub.com/building-a-home-security-system-with-Beaglebone/book) to integrate wireless technology with their existing wired system. I won't go too much into the details here except to say that it is basically the same circuit with connectors added to match the ones on the wired system.

The BeagleBone adapter board is slightly different, because it is designed to mount on the BeagleBone itself. The XBee breakout board has the same LEDs and switches connected to it. However, it gets its power from the BeagleBone and is connected to **UART 1** of the BeagleBone. The following is a circuit diagram of the BeagleBone-to-XBee adapter PCB:

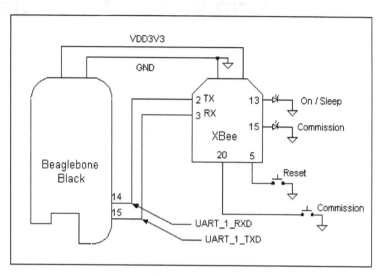

BeagleBone XBee adapter

The following image is of the prototype adapter board I built while writing this book. An actual PCB of the adapter is available for download along with the other files.

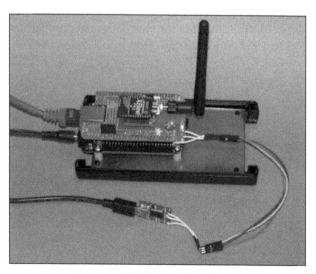

Prototype BeagleBone XBee adapter

For those of you who didn't buy my first book, *Building a Home Alarm System with BeagleBone*, I will provide a brief description of how the zone input board works, how it is constructed, and how you can test it before installing it.

Part 2B – the alarm system hardware

First of all, what is a zone? A **zone** is an area that is being protected and/or a collection of sensors that are protecting the zone. For example, ground floor windows could be a zone. Ground floor PIRs could be another zone. There are also special zones, such as fire alarm zones. In a commercial installation, fire zones and **carbon monoxide** (**CO**) detectors must have their own zone.

But first of all, the alarm system, in this case the BeagleBone, must be able to monitor the world around it. It does this by using ICs called comparators. You can think of a comparator as an op-amp with extremely high gain. Like an op-amp, when the voltage on the positive input (+) is higher than the voltage on the negative input (-), the output goes high. When the opposite is true, the output goes low.

We will be using an LM339 comparator. This device has an open collector output. What this means is that a resistor is required to pull the output high. By connecting the output pull-up resistor to 3.3 V, we now have a level converter that can be safely connected to the BeagleBone.

In the following diagram, a 4.7 K ohm resistor R5 is used as a pull up. The LM339 can sink plenty of current; so, for debugging purposes, we will also connect **D1** and **R6** to the output. That way when the comparator is triggered, D1 will turn on. The other handy feature of open collector outputs is that you can "OR" them together.

The following circuit actually monitors for two different conditions. If condition 1 is true then pin 2 goes low or if condition 2 is true, pin 1 goes low.

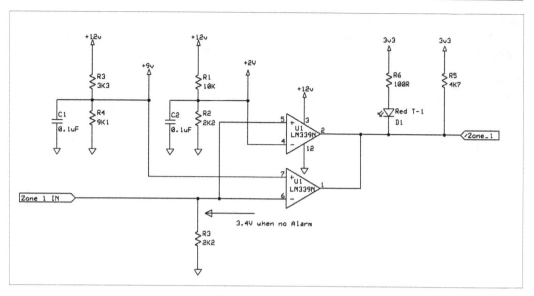

Alarm zone input circuit

So how does the circuit work?

What is an **End of Line (EOL)** resistor and what is it used for?

An EOL resistor is a resistor that is installed in the last device (sensor) of a chain of devices; hence the term "end of line resistor."

The resistors, **R4** and **R3**, and the EOL resistor form a voltage divider. Basically, you have 5.6K Ohms and 2.2K Ohms in series and connected to a 12-volt battery. Therefore, the ohms law tells us that there will be about 3.4 volts at the plus input of the comparator when the alarm contacts are closed.

Resistors **R1** and **R2** form a voltage divider that provides a 2-volt reference for the negative input. As we learned before, if the plus input (3.4V) is higher than the minus input (2.0V) then the output will be high.

If an alarm condition happens and the switch opens, then the 2.2K Ohms resistor, R3, will pull the input to 0 volts and the output will go low, because now it is less than 2.0V.

This also happens if the wire is cut by a burglar.

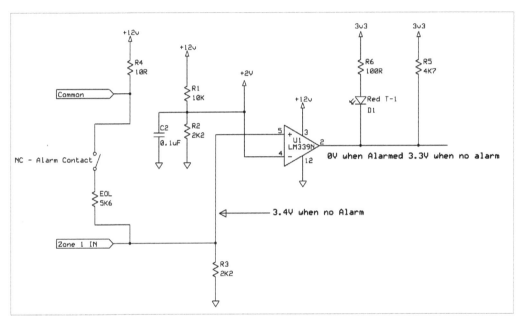

Open circuit alarm condition

But what if the burglar tries to jumper the alarm contacts and shorts out the EOL resistor?

In this case, the negative input of the LM339 goes higher than the positive input and once again, the output is pulled low. In this case, the reference voltage (9.0 V) is applied to the positive input and the sense voltage is applied to the minus input. If the EOL resistor is shorted, then 12 V is applied to the negative input of the comparator, and the output is pulled low. As we learned before, the outputs of the two circuits are ORed together, so that, if either an open OR a short occurs, the alarm will be triggered.

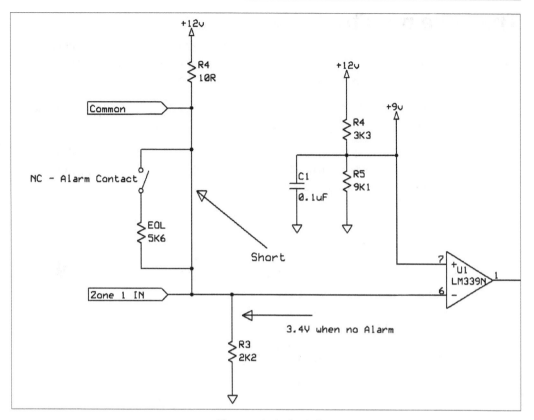

Alarm contacts shorted

You may have also noticed that there is a considerable difference between the sense voltage and the reference voltage. This is to allow for long runs of wire, for those of you who like to use Google, it is called hysteresis.

If you have a professionally installed alarm panel or know someone who does, open the panel and look at the terminal blocks where the wires are attached to the panel. If you see resistors connected across screw terminals with wires connected to the alarm system, this means it was not installed properly! What you do about this is up to you. They are called EOL resistors for a reason. They belong at the end of the line. They should each be installed on the last sensor in the loop.

The first is the module that monitors the various zones, in this case two zones. The second module is the optically isolated output board. This is a board that is used to activate lights and sirens and such. You can control up to four devices with this board.

The zone monitor PCB

This is a relatively simple board to assemble. All the parts used in this project are through-hole parts, so no advanced surface mount soldering skills are required. You will also find that this is much easier to test and repair.

When I assemble PCBs, I always start with the lowest parts and work towards the highest. That way, when you flip the board on its back, all the parts don't fall out!

Just follow the parts list and the silkscreen on the board and you should not have any problems. Just be sure to double-check that the integrated circuit U1 is installed correctly. The LEDs and diodes are also polarity sensitive.

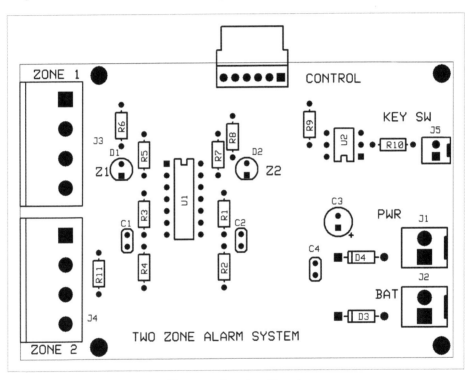

Two-zone alarm input board

The schematic for the two-zone alarm system is shown in the following diagram for reference purposes:

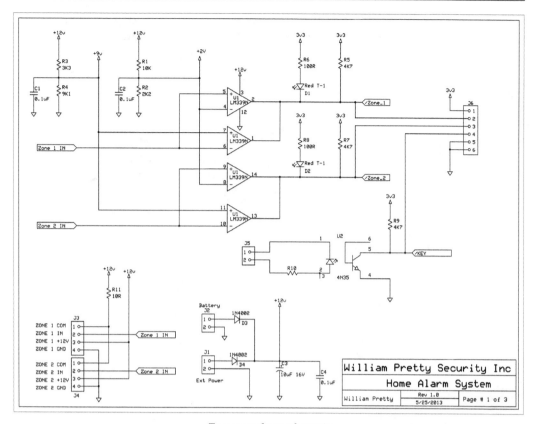

Two zone alarm schematic

The connectors **J3** and **J4** on the zone monitor schematic are your connections to the zones that you wish to monitor. Pins 1 and 2 of each connector are the zone inputs. Pins 3 and 4 are used to supply power to PIRs and other active sensors. Pin 4, which is called GND or ground, is the +12 V return. It is not the same as the COM or common input.

Connector **J5** is the key switch input. Applying +12 V to pin 2 and ground to pin 1 will turn on the opto-isolator **U2**. This will pull the **/key** signal low.

This +12 V can come from a key switch, a push button switch, a toggle switch, or anything you like as long as it applies +12 V to the opto-isolator.

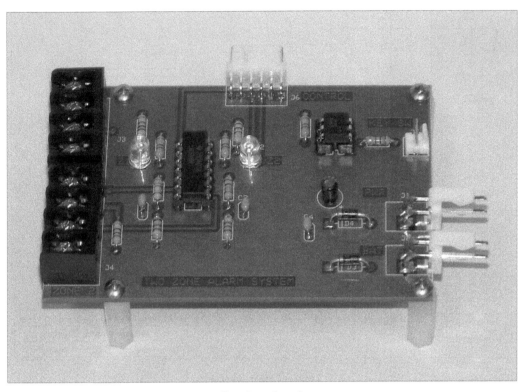

Zone input board

The preceding image shows the completed PCB. You will notice that I have used sockets for the ICs. This is not strictly necessary, but when I build prototypes, I use sockets until I am sure that the design will work. Also notice that the rectifier diodes **D3** and **D4** are raised off of the board. This was to allow better cooling, should they need it.

Isolated output PCB

The optically-isolated output PCB is a very simple construction. There are two reasons why I made a separate board in this case, as follows:

- This form factor allows a more modular design. You can have as many outputs as you can find I/O pins on your XBee modules.

- It is also cheaper to order small 3.8-inch x 2.5-inch mini boards from ExpressPCB than to order a larger custom-size board.

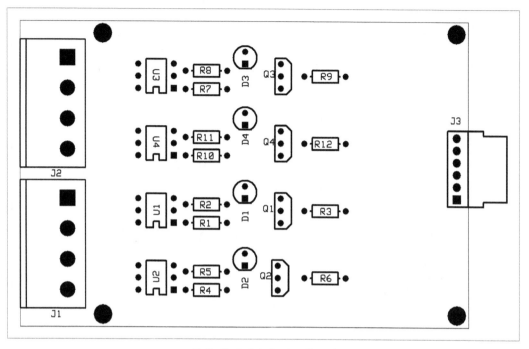

Optically isolated output board

The assembly of this board is very much the same as with the first board:

1. Install the resistors first.
2. Install the LEDs, making sure that the polarity is correct.
3. Install the transistors, again checking for correct installation.
4. Install the ICs, preferably on sockets to make repairs easier.

5. Finally, install the connectors.

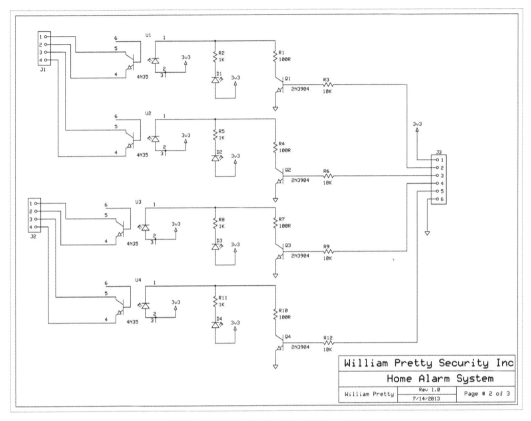

Optically isolated output board schematic

All four circuits are identical. A logical value 1 on the input applies 3.3 V to the base of the NPN transistor. This turns on the transistor, which pulls the collector of the transistor low. This turns on the LED and the opto-isolator. The LED is provided for debugging purposes. The LED turns on when the opto-isolator is activated.

Optically isolated output board

The preceding image shows the assembled PCB. Observant readers may notice that the color code of the 1K resistor for the first output is different from the rest. This resistor is what you're probably used to and has the normal 5-percent color code: brown, black, red, gold meaning 1 - 0 + 00 5-percent tolerance.

The other resistors are 1-percent tolerance resistors, so three decimal places are required: 1 - 0 - 0 + 0 1 percent.

There is no need to use 1-percent resistors in this design. I had them in my lab stock so I used them.

You will also note that the white connector hangs over the edge of the board. This is to make it easier to attach the connector when the boards are stacked during final assembly.

Connecting devices to the board

Congratulations! You now have a finished PCB capable of turning 3.3V logic signals into a real-world action. Not only can you turn on lights and sirens, you can also use this board to turn on your lawn sprinkler!

Here are two methods of doing just that:

- You can use a mechanical relay
- You can use a **solid state relay (SSR)**

The following diagram shows a small relay connected directly to the output transistor of the opto-isolator. If you want to maintain isolation between the alarm panel and the device you are controlling, the 12V battery powering the relay should have a separate ground return from the panel ground.

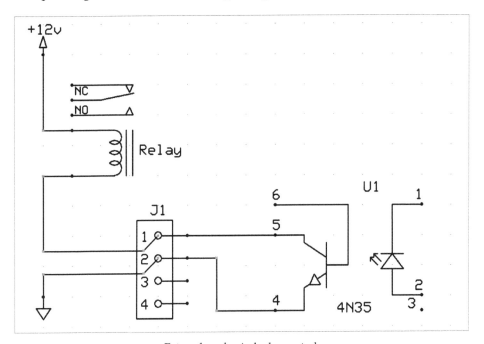

External mechanical relay control

You can now hook your siren or flashing light to the contacts of the mechanical relay without the fear of blowing the output transistor of the isolator on the board. Accidents happen, and that is why I suggested putting the opto-isolators on sockets. They aren't very expensive so I suggest you order spares.

In this case, we are using one opto-isolator to drive another. In the schematic form, this probably looks a bit odd. However, in the real world, the second isolator can be a high-current device capable of switching high voltage and current. In addition to the transistor shown in the drawing, it is also possible to buy opto-isolators with triac outputs for switching AC loads.

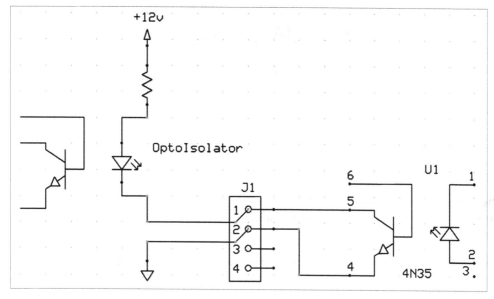

External opto-isolator connection

An example of one such AC load might be the motor that opens and closes your garage door or the gate at the end of your lane.

 The wiring of such high-voltage and high-current loads should probably be left to your local licensed electrician. Meeting the local electrical codes is up to you. As the system was built by you, it will *not* be CSA/UL approved, and the electrician may not want to install or connect it.

Another application of this type of relay would be to turn on exterior lighting. This could be done, either automatically using a zone input and a light sensor, or remotely by connecting to the panel via the Internet.

Solid state relay

The preceding image shows a high current SSR capable of switching AC loads. Note the two mounting fins on either side, so that the device can be mounted to a heat sink.

One thing I should mention about SSRs are that, in this project, the panel will be doubly isolated from the AC load. This is because you have one SSR driving another.

One handy use of an AC relay might be to turn on a backup sump pump when the alarm system detects a flood and the primary pump has failed. This would be handy for those of us who live in low-lying areas.

Part 3 – building the actual network (software)

First of all, we have to install the software that will be required to communicate with the XBee modules.

Step 1 – installing Oracle's JDK on BeagleBone Black

The software we will be writing requires Java and JavaScript, so the first step is to install Oracle's JDK. Then download the Linux ARM v6/v7 Hard Float ABI Version:

1. Copy `jdk-7u40-linux-arm-vfp-*.gz` to your BeagleBone Black.

2. Perform `tar xzf jdk-7u40-linux-arm-vfp-*.gz` to extract the JDK.

3. Perform `export PATH=$PATH:/home/root/jdk1.7.0_40/bin` to add the JDK to your path.

4. Perform `export JAVA_HOME=/home/root/jdk1.7.0_40` to set the `JAVA_HOME` on your installation.

5. Perform `java -version` to verify your installation.

Step 2 – installing the Java serial port extension package

To install this package, perform the following commands on your terminal:

```
apt-get install build-essential
sudo apt-get install nodejs
npm install serialport
```

Step 3 – enabling serial ports on startup

In the `serials.sh` file, use the following code:

```
#!/bin/bash
cd /lib/firmware/
echo ttyO1_armhf.com > /sys/devices/bone_capemgr*/slots
echo ttyO2_armhf.com > /sys/devices/bone_capemgr*/slots
echo ttyO4_armhf.com > /sys/devices/bone_capemgr*/slots
cd ~
```

Modify `rc.local` to run scripts on startup. Scripts should be in `/usr/local/bin/cd/etc`:

```
nano rc.local
!/bin/sh -e
#
# rc.local
#
# This script is executed at the end of each multiuser runlevel.
# Make sure that the script will "exit 0" on success or any other
# value on error.
#
# In order to enable or disable this script just change the execution
# bits.
#
# By default this script does nothing.

/usr/local/bin/serials.sh

exit 0
```

Testing the hardware

To test the hardware, perform the following steps:

1. The first thing we are going to do is run some code in the Cloud9 IDE, which will make sure that our network is configured correctly and that the hardware is working properly. We first have to power up our remote modules. We do this by applying about 6 volts to the DC input pin of our remote module board. Four AA cells in a holder will work fine for now.

2. You should see the LED connected to the SLEEP pin on continuously because we will not be using the sleep mode in this version of the software. The LED connected to the COMMISSION pin should be flashing, which indicates that it is ready to receive network packets.

3. Next, power up the BeagleBone, and wait for it to boot Debian. Once the boot sequence has finished, you can log in as root with the password `password`. The reason that we have to log in as root is that there appears to be a bug in the Cloud9 IDE that requires this method of login.

4. To find the IP Address that has been assigned to your BeagleBone, enter `ifconfig`.

5. To run the Cloud9 IDE, enter `cloud9.sh -l 0.0.0.0`.

6. Now that you have the IP Address, open a browser and enter the address plus port number 3131. For example, in my case, it was 192.168.10.105:3131.

The code we want to run is called send_packet_test.js and will send a command to all the remote XBee modules to turn on D1. This is the same function that is used in the actual alarm code. It has just been simplified for this test.

The first thing we have to do is to set up the serial port configuration using the JavaScript add-on we installed earlier. The baud rate is set to the same as the one we configured earlier using XCTU:

```
// Serial Port Setup
var serialPort, sp;
var comPort = '/dev/ttyO1';
var serialPort = require('serialport').SerialPort
sp = new serialPort(comPort, { baudrate: 9600,});
```

7. Now that we have the port configured, we need to set up some constants and variables:

```
// Constants and Variables
const frameStartByte = 0x7E;
const frameTypeRemoteAT   = 0x17;
const remoteATOptionApplyChanges = 0x02;


var sum;              // Calculated Checksum
var pin_state;        // On or Off 0x5=On 0x4=OFF
```

These variables will be used by the actual alarm software. They are hardcoded for this example:

```
var pin_num;          // Pin number 1 - 3
var mod_address_H;    // 16bit address of module High byte
var mod_address_L;    // 16bit address of module Low byte
```

8. Now we call the send_packet() function:

```
send_packet();


function send_packet(){
  sp.on("open", function () {
    console.log(comPort + ' is open');
    sum = 0;                 // Calculate the checksum as we go
```

What we are going to do is send the following packets to the other modules on the network:

To turn the LED on, we send these packets:

`7E 00 10 17 01 00 00 00 00 00 00 FF FE FF FE 02 44 31 05 71`

To turn the LED off, we send the following packets:

`7E 00 10 17 01 00 00 00 00 00 00 FF FE FF FE 02 44 31 04 72`

To do this we send a remote AT command that consists of an ASCII letter D followed by the ASCII number of the I/O port; in this case port D1:

```
/The text of the AT command
  sp.write([0x44]);
  sum += (0x44);       //Capital 'D'
  sp.write([0x31]);
  sum += (0x31);       //Number '1'
```

The actual value that we write is either 0x4 to turn the LED off or 0x5 to turn it on:

```
// The value (0x4 for off, 0x5 for on)
//sp.write[pin_state];
//sum +=(pin_state);
  sp.write([0x4]);
  sum +=(0x4);
```

Now we calculate the actual checksum based on the information we have been accumulating as we sent the bytes. Then all we have to do is send it and an LED should turn on at the other end:

```
    // Send the checksum
    var checksum = (0xFF - ( sum & 0xFF));
    // Mask off the MSB and then subtract from 0xFF to get checksum
       sp.write([checksum]);
       });
     }
```

The next feature we have to test is called I/O Line Passing. This feature is only available on series 1 XBee modules, and that is why they were chosen for this book. This feature vastly simplifies the software. The way line passing works is that an I/O output pin on the Coordinator is paired with I/O input pins on the end units. So, if a pin on one of the end (remote) units goes low, the corresponding pin on the Coordinator goes low. All we have to do is monitor the output pin of the Coordinator XBee module with the BeagleBone and trigger an alarm when something happens.

That is exactly what the following JavaScript does.

The first step is to get the bonescript add-on for JavaScript and then enable the output pins on the BeagleBone:

```javascript
var b = require('bonescript');

var outputPin1 = "P8_13";        //Alarm Output 1
var outputPin2 = 'P8_11';        //Alarm Output 2
var outputPin3 = 'P8_12';        //Alarm Output 3
var outputPin4 = 'P8_14';        //Alarm Output 4
var outputPin5 = 'P9_27';        //Xbee Alarm Output

// configure output pins and set all low
b.pinMode(outputPin1, 'out');
b.pinMode(outputPin2, 'out');
b.pinMode(outputPin3, 'out');
b.pinMode(outputPin4, 'out');
b.pinMode(outputPin5, 'out');

//Setup input pin directions
b.pinMode(inputPin1, b.INPUT);
b.pinMode(inputPin2, b.INPUT);
b.pinMode(inputPin3, b.INPUT);
b.pinMode(inputPin4, b.INPUT);
b.pinMode(xBeePin1, b.INPUT);
b.pinMode(xBeePin2, b.INPUT);
b.pinMode(keyPin, b.INPUT);

// This is where the Alarm System code starts

var inputPin1 = 'P8_19';        //Zone 1 is connected to this pin
var inputPin2 = 'P8_15';        //Zone 2 is connected to this pin
var inputPin3 = 'P8_9';         //Zone 3 is connected to this pin
var inputPin4 = 'P8_7';         //Zone 4 is connected to this pin
var keyPin = 'P8_17';           //Key switch connected to this pin
var xBeePin1 = 'P9_15';         //Remote Zone 1 Alarm
var xBeePin2 = 'P9_23';         //Remote Zone 2 Alarm
```

Next, we have to attach an interrupt to each input pin so that an interrupt handler will be called if the pin goes low:

```
// Setup interupts
b.attachInterrupt(inputPin1, true, b.FALLING, Zone1Callback);
b.attachInterrupt(inputPin2, true, b.FALLING, Zone2Callback);
b.attachInterrupt(inputPin3, true, b.FALLING, Zone3Callback);
b.attachInterrupt(inputPin4, true, b.FALLING, Zone4Callback);
b.attachInterrupt(xBeePin1, true, b.FALLING, XBeeZone1Callback);
b.attachInterrupt(xBeePin2, true, b.FALLING, XBeeZone2Callback);
b.attachInterrupt(keyPin, true, b.FALLING, keyCallback);
```

Now that we have our interrupts set up, all we have to do is sit back and wait for something to happen. When it does, we send a message to the console and then turn on the opto-isolator to signal an alarm:

```
console.log ('Waiting for an Alarm Interupt');

//Alarm Interupt Handlers Start Here:
//Alarm Detected in Remote Zone 1
function XBeeZone1Callback() {
  flag4 ++;
  if (flag4 > 1){
    console.log ("Alarm Detected in Zone 4");
    state = 1;
    b.digitalWrite(outputPin5, state);//Turn ON the Opto-Isolator
  }
}


//Alarm Detected in Remote Zone 1
function XBeeZone1Callback() {
  flag4 ++;
  if(flag4 > 1){
    console.log ("Alarm Detected in Zone 4");
    state = 1;
    b.digitalWrite(outputPin5, state);//Turn ON the Opto-Isolator
  }
}
```

If you got this far and everything is working, congratulations! You now are the proud owner of a wireless alarm system.

For the more adventurous among you, I have included a *Going further* section.

Going further

In this section, I will describe to you the makeup of the packet that the end unit sends to the coordinator. This is important for you software folks, because it will tell you which remote or end unit triggered the alarm; assuming that you have more than one end unit.

The following is a sample of packets received from two different end units:

```
Module #1

7E 00 0A 83 00 00 3F 00 01 00 06 00 04 32 <- Button 1 Down
7E 00 0A 83 00 00 41 00 01 00 06 00 06 2E <- Button 1 Up


7E 00 0A 83 00 00 4B 00 01 00 06 00 02 28 <- Button 2 Down
7E 00 0A 83 00 00 53 00 01 00 06 00 06 1C <- Button 2 Up


===========================================


Module #2

7E 00 0A 83 00 00 30 00 01 00 06 00 04 41
7E 00 0A 83 00 00 31 00 01 00 06 00 06 3E


7E 00 0A 83 00 00 32 00 01 00 06 00 02 41
7E 00 0A 83 00 00 34 00 01 00 06 00 06 3B
```

The following is an explanation of the packet contents:

```
Module #1

7E 00 0A 83 00 00 3F 00 01 00 06 00 04 32 <- Button 1 Down
7E 00 0A 83 00 00 41 00 01 00 06 00 06 2E <- Button 1 Up
```

The following table is a breakdown of the packet contents and what the various bytes mean:

Field Name	Module#1 Button Down	Description
Start Delimiter	7E	All API frames start with 7E.
Length	000A	Length of the frame (not including Checksum).
Frame Type	83	This is a 16-bit I/O sample.

Field Name	Module#1 Button Down	Description
16-bit Source Address	0000 *	16-bit address of the XBee that sent the sample. This matches the MY of the remote module.
RSSI	3F	Signal strength in dBm of the received packet. (-63dBm).
Option Byte	00	Additional information about the frame.
Number of Samples	01	How many samples are included, this is determined by the IT parameter.
Channel Mask	00 06	A bit field mask that indicates which analog and digital lines are configured as inputs.
Digital Samples	00 04	Digital sample data.
Analog Sample #1	**	Two-byte analog sample.
Analog Sample #2	**	Two-byte analog sample.
Analog Sample #3	**	Two-byte analog sample.
Checksum	32	Checksum of the API frame.

Note the following:

- * This is the 16-bit address. I left it blank; the user should assign an address so that they can identify the module that is sending the information. That way, all you have to do is parse the first few bytes of the frame.
- ** If no analog channels are enabled, then these six bytes are not sent.

If we take a look at the two different digital samples 00 04 and 00 06, they indicate that DIO1 has gone low (04) and then returned high (06).

The following is another useful table as a reference:

BeagleBone UART pins	RX	TX	CTS	RTS	Device	Remarks
UART0	J1_4	J1_5	-	-	/dev/ttyO0	For BeagleBone Black only
UART1	P9_26	P9_24	P9_20	P9_19	/dev/ttyO1	
UART2	P9_22	P9_21	P8_37	P8_38	/dev/ttyO2	
UART3	-	P9_42	P8_36	P8_34	/dev/ttyO3	TX only
UART4	P9_11	P9_13	P8_35	P8_33	/dev/ttyO4	
UART5	P8_38	P8_37	P8_31	P8_32	/dev/ttyO5	

Essentially, what you do is read an analog input and convert the reading to millivolts by dividing the reading by 4096 (the maximum reading from the ADC) and then multiplying by 1800 (the reference voltage). This will give you the actual reading in millivolts.

Now that we have the reading in millivolts, it is time to convert the reading to degrees Celsius. To do this we subtract 500 mV (the output offset of the TMP36) from the calculation and then divide by 10. We now have the reading in degrees Celsius.

For those of you who live in countries that still use the Fahrenheit scale, simply multiply the Celsius reading by 9, divide it by 5, and add 32, just like they taught you in school.

Summary

Well folks, that's it for this book. I hope you have enjoyed reading it as much as I did writing it. By now you have discovered that this is not your average BeagleBone projects book. I have tried to include something for everyone in this book.

This chapter was by far the most advanced. In it, we built our own wireless alarm system, using ZigBee technology and our very own alarm system boards.

Index

Cloud9 104
Device Tree Compiler 101
Device Tree Source 101
Node.js 102
profiles in root, modifying 104
socket.io 104
steps 100, 101
devices
connecting, to board 142-144
Device Tree Compiler
installing 101
drivers
URL 116
DUT (Device Under Test) 59

E

End of Line (EOL) resistor 133
end unit 124

F

FET (Field Effect Transistor) 58
filesystem
expanding 98, 99
filter
need for 64
filtering
results 65
flasher PCB
assembling 24, 25
flashlight
about 12-16
plastic lens, removing from 12, 13
floodlights
turning on 39

H

hardware
testing 146-150
hardware setup 95
headlamp 17, 18
HTML code 112, 113

I

impedance
URL 60
input impedance
URL 45
input section, sound cards 44-47
Integrated Development
Environment (IDE) 105
isolated output PCB 138-141

J

JavaScript 109-111
Java serial port extension package
installing 145
jQuery
URL 78
jQuery Mobile
URL 78

L

latch enable (LE) 76
Linear Technology
URL 84
log-periodic antenna
URL 90

M

mercury switch
working 35
mercury tilt switch 35
mesh networking
URL 115
model, Creative Labs
specifications 41, 42
URL 41
monostable mode
555 timer, working in 31
multimeter 10

Thank you for buying
Getting Started with Electronic Projects

About Packt Publishing

Packt, pronounced 'packed', published its first book, *Mastering phpMyAdmin for Effective MySQL Management*, in April 2004, and subsequently continued to specialize in publishing highly focused books on specific technologies and solutions.

Our books and publications share the experiences of your fellow IT professionals in adapting and customizing today's systems, applications, and frameworks. Our solution-based books give you the knowledge and power to customize the software and technologies you're using to get the job done. Packt books are more specific and less general than the IT books you have seen in the past. Our unique business model allows us to bring you more focused information, giving you more of what you need to know, and less of what you don't.

Packt is a modern yet unique publishing company that focuses on producing quality, cutting-edge books for communities of developers, administrators, and newbies alike. For more information, please visit our website at www.packtpub.com.

About Packt Open Source

In 2010, Packt launched two new brands, Packt Open Source and Packt Enterprise, in order to continue its focus on specialization. This book is part of the Packt Open Source brand, home to books published on software built around open source licenses, and offering information to anybody from advanced developers to budding web designers. The Open Source brand also runs Packt's Open Source Royalty Scheme, by which Packt gives a royalty to each open source project about whose software a book is sold.

Writing for Packt

We welcome all inquiries from people who are interested in authoring. Book proposals should be sent to author@packtpub.com. If your book idea is still at an early stage and you would like to discuss it first before writing a formal book proposal, then please contact us; one of our commissioning editors will get in touch with you.

We're not just looking for published authors; if you have strong technical skills but no writing experience, our experienced editors can help you develop a writing career, or simply get some additional reward for your expertise.

Raspberry Pi Cookbook for Python Programmers

ISBN: 978-1-84969-662-3   Paperback: 402 pages

Over 50 easy-to-comprehend tailor-made recipes to get the most out of the Raspberry Pi and unleash its huge potential using Python

1. Install your first operating system, share files over the network, and run programs remotely.

2. Unleash the hidden potential of Raspberry Pi's powerful VideoCore IV graphics processor with your own hardware accelerated 3D graphics.

3. Discover how to create your own electronic circuits to interact with Raspberry Pi.

BeagleBone Home Automation

ISBN: 978-1-78328-573-0   Paperback: 178 pages

Live your sophisticated dream with home automation using BeagleBone

1. Practical approach to home automation using BeagleBone; starting from the very basics of GPIO control and progressing up to building a complete home automation solution.

2. Covers the operating principles of a range of useful environment sensors, including their programming and integration to the server application.

3. Easy-to-follow approach with electronics schematics, wiring diagrams, and controller code all broken down into manageable and easy-to-understand sections.

Please check **www.PacktPub.com** for information on our titles

Building a Home Security System with BeagleBone

ISBN: 978-1-78355-960-2   Paperback: 120 pages

Build your own high-tech alarm system at a fraction of the cost

1. Build your own state-of-the-art security system.

2. Monitor your system from anywhere you can receive e-mail.

3. Add control of other systems such as sprinklers and gates.

4. Save thousands on monitoring and rental fees.

BeagleBone Robotic Projects

ISBN: 978-1-78355-932-9   Paperback: 244 pages

Create complex and exciting robotic projects with the BeagleBone Black

1. Get to grips with robotic systems.

2. Communicate with your robot and teach it to detect and respond to its environment.

3. Develop walking, rolling, swimming, and flying robots.

Please check **www.PacktPub.com** for information on our titles